INTERNATIONAL SERIES OF MONOGRAPHS IN

NATURAL PHILOSOPHY

GENERAL EDITOR: D. TER HAAR

VOLUME 27

INTRODUCTION TO QUANTUM ELECTRONICS

INTRODUCTION TO QUANTUM ELECTRONICS

BY

H. G. UNGER

Technical University of Braunschweig, Germany

PERGAMON PRESS

OXFORD · LONDON · EDINBURGH · NEW YORK

TORONTO · SYDNEY · PARIS · BRAUNSCHWEIG

Pergamon Press Ltd., Headington Hill Hall, Oxford
4 & 5 Fitzroy Square, London W.1
Pergamon Press (Scotland) Ltd., 2 & 3 Teviot Place, Edinburgh 1
Pergamon Press Inc., Maxwell House, Fairview Park, Elmsford, New York 10523
Pergamon of Canada Ltd., 207 Queen's Quay West, Toronto, Ontario
Pergamon Press (Aust.) Pty. Ltd., 19a Boundary Street, Rushcutters Bay, N.S.W. 2011, Australia
Pergamon Press S.A.R.L., 24 rue des Écoles, Paris 5e
Vieweg & Sohn GmbH, Burgplatz 1, Braunschweig

First Edition 1970

Library of Congress Catalog Card No. 76-86534

Printed in Hungary

08 006368 3

Contents

Contents

Preface

THIS small book is the text of a one semester lecture course which was originally prepared in German for students of electrical engineering and presented for the first time during the summer term of 1966 at the Technische Hochschule, Braunschweig, Germany. It is by no measure a complete exposition of quantum electronics which by now includes a rather wide and detailed range of topics. It intends rather to give only an introduction to the fundamentals of lasers and masers and stresses the physical principles, their theory, and methods of analysis in order to describe all relevant phenomena also quantitatively. Not much consideration has been given to technological aspects and practical results.

The text is intended for electrical engineers and applied physicists and aims to prepare them for the many specialized and advanced publications in quantum electronics.

Only a basic knowledge of the fundamentals of physics and electrical engineering is required to understand this text. However, all quantum mechanical methods of analysis as well as the underlying electromagnetic theory are given only in compact and brief form. The reader will therefore find it easier to follow the respective chapters if he is familiar with quantum mechanics and electromagnetic theory.

For a more detailed and advanced study of quantum electronics and related topics books and review articles are listed in the first part of the bibliography.

The author is indebted to Dr. D. M. Bolle of Brown University who reviewed the entire manuscript and made extensive suggestions.

Introduction

ELECTRONICS, however large an area of engineering science and technology, is still part of electrical engineering. Quantum electronics, therefore, must also be regarded as a branch of the engineering sciences. To be even more specific, it is a special branch of electronics within the field of electrical engineering. In the most general sense it should be concerned with all the applications of physical effects in electrical engineering in which the quantum nature of matter and fields as described by quantum mechanics and quantum electrodynamics is of any consequence. However, common practice has been to consider quantum electronics as being the technical application of the "so-called" induced or stimulated emission only. The utilization of this physical phenomenon has led to the two new electronic devices, the *laser* and *maser*. According to their mechanism both are intimately related, despite the difference in their respective practical applications.

Laser is an acronym for *l*ight *a*mplification by *s*timulated *e*mission of *r*adiation. Maser is the corresponding acronym for *m*icrowave *a*mplification. . . . The difference therefore lies first of all in the range of frequencies in which these devices amplify electromagnetic waves.

The maser, which was first conceived and operated in the nineteen-fifties, amplifies in a frequency range for which there had existed efficient amplifiers long before. It was, however, much more sensitive than the previous microwave amplifiers known at that time. Maser noise was nearly two orders of magnitude smaller at the time than the noise figure for low noise travelling wave tubes or sensitive converter stages. The maser is indeed so low in noise that its application is only justified where signals must be detected from very quiet or cold surroundings. These conditions prevail for communication satellite ground stations or radio astronomy.

More recently other microwave amplifiers have also been made so sensitive that they compete with masers even in these applications and are beginning to replace them. Parametric amplifiers, in particular those which use *p–n* junction varactors and which are cooled with liquid helium, are nearly as low in noise as the most sensitive masers. The laser was first developed and operated in 1960. In contrast to the limited application of masers in an established technology, the laser opened up a wide spectrum of electromagnetic waves for new and revolutionary applications. The range of this spectrum extends from submillimetre waves up into the ultraviolet frequencies including infrared and visible light.

The significance of their intensity, or frequency, or both, depends on the way in which the waves, out of this wide spectral range, are observed or utilized. It may also be important that they extend over well-defined limits in time and space or that they are generated coherently over certain regions in time or space. All these characteristics, which have been perfected close to the ultimate limits for electromagnetic waves of lower frequencies, may now be enormously improved through quantum mechanical amplifiers at these higher frequencies also.

Even when these amplifiers are only made to generate coherent optical frequencies by self-oscillation through positive feedback most of these improvements become evident. Previously electromagnetic waves could be generated coherently only at frequencies lower than 10^{12} Hz corresponding to a shortest wavelength of 0·3 mm. Waves can only be focused to within the diffraction limit which in turn is determined by their wavelength. Moreover, only modest power could previously be obtained near the high-frequency limit. Thus the intensity, even with optimum focusing, was severely limited.

It is true that high resolution and intensity have been obtained before at optical frequencies, but they could only be generated incoherently. Even when coherence is of no direct significance the focusing power and intensity of ordinary incoherent light sources are much lower than the laws of diffraction would allow for coherent light. The atomic oscillators that generate ordinary light radiate independently from each other with random phase and polarization. By forming the image of such a source the highest intensity that can be obtained is that of the light source itself.

Within the laser, however, many of the atomic oscillators radiate energy at one and the same frequency and in phase with an exciting wave. Atomic oscillators in a large volume of active material radiate in synchronism. They amplify or generate a harmonic oscillation which throughout the volume of the medium has the well-defined field configuration of a continuous wave and which forms a coherent wave train as long as the coherent emission lasts. Thus the electromagnetic oscillation and the wave which is radiated are coherent in time and space.

All physical experiments and technical applications of electromagnetic waves and oscillations, which depended on high coherence, had to be confined previously to frequencies of less than 10^{12} Hz. For these experiments and applications the laser now extends the full electromagnetic spectrum into the far ultraviolet. All limitations set by frequency have now been increased by at least a factor of 10^3.

Even when coherence is of no direct significance, but it is important that the intensity be high or that the radiation be sharply focused in space or time, coherent waves will be much more efficient than incoherent waves. Focusing power is now only diffraction limited, thus allowing the rather high but still limited radiation density of a large laser to be focused to extremely high intensities. To control the extension of coherent waves in time precisely they can be switched electronically or mechanically much faster than incoherent waves.

CHAPTER 1

Stimulated Emission and Absorption

1.1. A simple classical model

The mechanism of amplification in lasers and masers is based on radiation of energy from an atomic oscillator system induced by external electromagnetic fields. This phenomenon can only be described satisfactorily through the methods of quantum mechanics. Nevertheless, elementary classical treatment will help to understand the subsequent results obtained from quantum mechanics.

In this classical treatment an electron is assumed to oscillate with angular frequency ω within a potential well of harmonic shape. Such a harmonic oscillator well may be considered an approximate model of the force field of the atomic system acting upon the electron. Let the electron's displacement from its zero location be

$$\mathbf{r} = \mathbf{r}_0 \sin(\omega t + \phi) \tag{1}$$

The oscillating electron is subject to an external electric field

$$\mathbf{E} = \sqrt{2}\, \mathbf{E}_0 \cos \omega t \tag{2}$$

Depending on the orientation of the field and the phase ϕ between the electron oscillation and the field the electronic oscillation will either be amplified or damped. Therefore, energy is either absorbed by the electron from the field or supplied by the electron to the field. An instantaneous time constant is defined by the rate of energy exchange between electron and field. If $\mathbf{E}_0$ and $\mathbf{r}_0$ are parallel in space this instantaneous time constant is

$$\tau = \frac{m\omega r_0}{\sqrt{2}eE_0 \cos \phi} \tag{3}$$

where m and e are the mass and charge of the electron.

In the range $-\pi/2 < \phi < \pi/2$ we have $\tau > 0$ and the total energy of the electron is decreasing. Emission is taking place in this range of ϕ. For the remaining range of ϕ we have $\tau < 0$ and absorption of energy by the electron. Both processes are induced by the external field. They are therefore called induced or stimulated emission and absorption respectively.

If the electric field acts upon a large number of electrons oscillating in their own potential wells but with random phase and orientation, emission and absorption will be equal on the average and there is no net exchange of energy. Only if one phase and orientation prevail will there be a resultant induced emission or absorption. For induced emission to take place a dominant number of the atomic oscillators of the classical model must be in a suitable phase range.

The classical model can thus explain the exchange of energy. Accordingly if all states, i.e. phases ϕ of the oscillators, are equally populated, induced emission must be equal to induced absorption. Beyond that, however, no quantitative information can be obtained from this model. For a more detailed and quantitative description the methods of quantum mechanics must be used.

To proceed further it is first necessary to supplement the commonly known laws of classical mechanics by a few concepts of analytical mechanics. These concepts will be needed later in the quantum mechanical treatment.

1.2. The Hamiltonian function

The Hamiltonian function H and the Lagrangian function L are introduced into classical mechanics to obtain universal forms for the equations of motion. For so-called conservative systems these functions are

$$H = T+U, \quad L = T-U \tag{4}$$

Here T is the kinetic energy of the system and U its potential energy. Conservative systems are those in which only conservative forces are acting. For a point system with N points of mass these forces may be derived from the potential energy U as follows:

$$F_{in} = -\frac{\partial U}{\partial x_{in}}$$

Here x_{in} is the ith cartesian coordinate of the nth point of mass. U is a function of the position of all points of mass:

$$U = U(x_{in})$$

Newton's equations of motions are

$$m_n \ddot{x}_{in} = F_{in}$$

where m_n is the mass of point n and $\ddot{x}_{in}$ indicates the second derivative of x_{in} with respect to time, i.e. the acceleration in x_{in}.

In these equations of motion the forces may be expressed in terms of the potential energy

$$m_n \ddot{x}_{in} = -\frac{\partial U}{\partial x_{in}} \tag{5}$$

By eliminating U we obtain

$$H = 2T - L$$

from eqn. (4).

The kinetic energy in this expression may be written as

$$2T = \sum_{n=1}^{N} \sum_{i=1}^{3} m_n \dot{x}_{in}^2$$

with $\dot{x}_{in}$ the first derivative of x_{in} with respect to time, i.e. the velocity of point n in the x_{in}-direction.

To avoid being restricted to the rather specialized point system of masses and its representation in Cartesian coordinates, we introduce generalized coordinates of position $q_1, q_2, q_3, \ldots, q_n$. The transformation from Cartesian coordinates to these generalized coordinates is described by a system of $3N$ equations:

$$x_{in} = f_{in}(q_1, q_2, q_3, \ldots, q_{3N})$$

The differentials then become

$$dx_{in} = \sum_{k=1}^{3N} \frac{\partial x_{in}}{\partial q_k} dq_k \tag{6}$$

Employing this coordinate transformation the kinetic energy is

$$2T = \sum_{n=1}^{N} \sum_{i=1}^{3} m_n \left(\frac{\partial x_{in}}{\partial q_1} \dot{q}_1 + \frac{\partial x_{in}}{\partial q_2} \dot{q}_2 + \dots \right)^2 \tag{7}$$

where the dots on the q's again indicate the derivatives with respect to time.

In Cartesian coordinates the components of momentum are:

$$p_{in} = m_n \dot{x}_{in} = \frac{1}{2} \cdot \frac{\partial}{\partial \dot{x}_{in}} \left(\sum_m \sum_l m_m \dot{x}_{lm}^2 \right) = \frac{\partial T}{\partial \dot{x}_{in}} \tag{8}$$

In generalizing this expression the quantities

$$p_k = \frac{\partial T}{\partial \dot{q}_k} \tag{9}$$

are called the general components of momentum. Substituting from (7) these general components of momentum may also be written as

$$p_k = \sum_{n=1}^{N} \sum_{i=1}^{3} m_n \left(\frac{\partial x_{in}}{\partial q_1} \dot{q}_1 + \frac{\partial x_{in}}{\partial q_2} \dot{q}_2 + \dots \right) \frac{\partial x_{in}}{\partial q_k}$$

Considering eqn. (7) and this latter relation we obtain

$$\sum_{k=1}^{3N} \dot{q}_k p_k = 2T.$$

Therefore Hamilton's function may also be expressed as

$$H = \sum_{k=1}^{3N} \dot{q}_k p_k - L$$

The total differential of this expression is

$$dH = \sum_{k=1}^{3N} p_k \, d\dot{q}_k + \sum_{k=1}^{3N} \dot{q}_k \, dp_k - \sum_{k=1}^{3N} \frac{\partial L}{\partial \dot{q}_k} d\dot{q}_k - \sum_{k=1}^{3N} \frac{\partial L}{\partial q_k} dq_k$$

since L is a function of the general coordinates of position as well as of their time derivatives, the latter being the general components of velocity. In contrast the potential energy U is a function of position only and is independent of these velocity components. Therefore from

$$L = T - U$$

we obtain

$$\frac{\partial L}{\partial \dot{q}_k} = \frac{\partial T}{\partial \dot{q}_k} = p_k \tag{10}$$

Hence

$$dH = \sum_{k=1}^{3N} \dot{q}_k \, dp_k - \sum_{k=1}^{3N} \frac{\partial L}{\partial q_k} dq_k \tag{11}$$

The partial derivatives $\partial L/\partial q_k$ are directly related to the values p_k through the equations of motion. To obtain these relations we consider the equations of motion again in Cartesian coordinates and substitute from (8) for $\ddot{x}_{in}$:

$$\frac{d}{dt}\left(\frac{\partial T}{\partial \dot{x}_{in}}\right) + \frac{\partial U}{\partial x_{in}} = 0$$

To transform into general coordinates we multiply each of these equations by $\partial x_{in}/\partial q_k$ respectively, and sum over all i and n:

$$\frac{d}{dt} \sum_{i,n} \frac{\partial T}{\partial \dot{x}_{in}} \frac{\partial x_{in}}{\partial q_k} - \sum_{i,n} \frac{\partial T}{\partial \dot{x}_{in}} \frac{d}{dt}\left(\frac{\partial x_{in}}{\partial q_k}\right) + \sum_{i,n} \frac{\partial U}{\partial x_{in}} \frac{\partial x_{in}}{\partial q_k} = 0 \tag{12}$$

For the coordinate transformation we have according to eqn. (6)

$$\dot{x}_{in} = \sum_{k=1}^{3N} \frac{\partial x_{in}}{\partial q_k} \dot{q}_k \tag{13}$$

Also the following relations hold

$$\frac{\partial U}{\partial q_k} = \sum_{i,n} \frac{\partial U}{\partial x_{in}} \cdot \frac{\partial x_{in}}{\partial q_k}$$

$$\frac{\partial T}{\partial \dot{q}_k} = \sum_{i,n} \frac{\partial T}{\partial \dot{x}_{in}} \cdot \frac{\partial \dot{x}_{in}}{\partial \dot{q}_k}$$

$$\frac{\partial T}{\partial q_k} = \sum_{i,n} \frac{\partial T}{\partial \dot{x}_{in}} \cdot \frac{\partial \dot{x}_{in}}{\partial q_k}$$

From eqn. (13) follows

$$\frac{\partial \dot{x}_{in}}{\partial \dot{q}_k} = \frac{\partial x_{in}}{\partial q_k}$$

while we have

$$\frac{d}{dt}\left(\frac{\partial x_{in}}{\partial q_k}\right) = \sum_{l=1}^{3N} \frac{\partial}{\partial q_l}\left(\frac{\partial x_{in}}{\partial q_k}\right)\dot{q}_l = \frac{\partial}{\partial q_k}\sum_{l=1}^{3N}\frac{\partial x_{in}}{\partial q_l}\dot{q}_l = \frac{\partial \dot{x}_{in}}{\partial q_k}$$

From all these transformation relations the equations of motion in general coordinates are

$$\frac{d}{dt}\left(\frac{\partial T}{\partial \dot{q}_k}\right) - \frac{\partial T}{\partial q_k} + \frac{\partial U}{\partial q_k} = 0 \qquad (k = 1, 2, \ldots, 3N)$$

Using $L = T - U$ and (9) these equations reduce simply to

$$\frac{\partial L}{\partial q_k} = \dot{p}_k$$

When the equations of motion are taken into account the total differential (11) of Hamilton's function is given by

$$dH = \sum_{k=1}^{3N} (\dot{q}_k\, dp_k - \dot{p}_k\, dq_k)$$

Alternatively H may be considered a function of the general coordinates of position q_k and momentum p_k. Then the total differential is

$$dH = \sum_{k=1}^{3N} \frac{\partial H}{\partial p_k} dp_k + \sum_{k=1}^{3N} \frac{\partial H}{\partial q_k} dq_k$$

Comparing these two expressions they can only be identical for all dp_k and dq_k if the following system of equations holds

$$\left.\begin{aligned} \frac{\partial H}{\partial p_k} &= \dot{q}_k \\ \frac{\partial H}{\partial q_k} &= -\dot{p}_k \end{aligned}\right\} \quad (k = 1, 2, \ldots, 3N) \tag{14}$$

These are Hamilton's equations of motion. Because of their universal form they are often called canonical equations of motion. They have been derived here for conservative systems.

While Hamilton's equations have been obtained for systems with conservative forces only it has proved advantageous to employ them also in treating systems with non-conservative forces. In this case

Hamilton's equations are postulated to hold for the non-conservative system. Hamilton's function is here no longer defined by eqn. (4) but is now defined as that function which satisfies Hamilton's equations for the system.

1.3. Hamilton's function for charge carriers in an electromagnetic field

When considering induced emission and absorption the interaction between charge carriers and electromagnetic fields has to be analysed. In order to prepare for the quantum mechanical treatment of this problem we must first find Hamilton's function of the system.

The electromagnetic field will be described by its vector-potential **A** and scalar potential φ. In source free regions **A** is a solution of the vector wave equation

$$\nabla^2\mathbf{A}-\frac{1}{c^2}\frac{\partial^2\mathbf{A}}{\partial t^2}=0$$

and φ of the scalar wave equation

$$\nabla^2\varphi-\frac{1}{c^2}\frac{\partial^2\varphi}{\partial t^2}=0$$

where $c=1/\sqrt{\mu\varepsilon}$ is the velocity of plane wave propagation in the medium and μ and ε are its permeability and permittivity, respectively. φ and **A** are related by the Lorentz condition

$$(\nabla\cdot\mathbf{A})=-\frac{1}{c^2}\dot{\varphi}$$

The electric field vector **E** and the magnetic induction **B** or magnetic field **H** are derived from the potentials as follows

$$\mathbf{E}=-\frac{\partial\mathbf{A}}{\partial t}-\nabla\varphi;\quad \mathbf{B}=\mu\mathbf{H}=\nabla\times\mathbf{A}$$

A particle of charge Q and mass m which is moving with velocity **v** in the electromagnetic field experiences a Lorentz force

$$\mathbf{F}=Q\mathbf{E}+Q(\mathbf{v}\times\mathbf{B})$$

Therefore the equation of motion for this charge carrier is

$$m\dot{\mathbf{v}} = Q\mathbf{E}+Q(\mathbf{v}\times\mathbf{B}) \tag{15}$$

Hamilton's function for this system is stated to be

$$H(p_k, q_k) = \frac{1}{2m}(\mathbf{p}-Q\mathbf{A})^2+Q\varphi \tag{16}$$

where $\mathbf{p}$ is a vector formed by the three components p_k of linear momentum of the charge carrier.

We will now show that this statement is correct by substituting (16) for H into Hamilton's equation and thereby obtaining the equations of motion (15).

In the cartesian coordinate system (x, y, z) the x-components of Hamilton's equations are

$$\dot{x} = \frac{1}{m}(p_x-QA_x)$$

$$\dot{p}_x = -Q\frac{\partial\varphi}{\partial x}+\frac{Q}{m}\left[(p_x-QA_x)\frac{\partial A_x}{\partial x}+(p_y-QA_y)\frac{\partial A_y}{\partial x}+(p_z-QA_z)\frac{\partial A_z}{\partial x}\right]$$

Corresponding relations hold for the other two cartesian components. The equations for $\dot{x}$, $\dot{y}$ and $\dot{z}$ are all included in the following vector equation

$$\mathbf{v} = \frac{1}{m}(\mathbf{p}-Q\mathbf{A}) \tag{17}$$

while the equation for p_x, using these relations for $\dot{x}$, $\dot{y}$, $\dot{z}$, may be simplified as follows

$$\dot{p}_x = -Q\frac{\partial\varphi}{\partial x}+Q\left(\dot{x}\frac{\partial A_x}{\partial x}+\dot{y}\frac{\partial A_y}{\partial x}+\dot{z}\frac{\partial A_z}{\partial x}\right)$$

The total derivative of $A_x\,(x, y, z, t)$ with respect to time is

$$\frac{dA_x}{dt} = \frac{\partial A_x}{\partial t}+\dot{x}\frac{\partial A_x}{\partial x}+\dot{y}\frac{\partial A_x}{\partial y}+\dot{z}\frac{\partial A_x}{\partial z}$$

Subtracting Q times this equation from the preceding equation we obtain:

$$\frac{d}{dt}(p_x-QA_x) = -Q\left(\frac{\partial\varphi}{\partial x}+\frac{\partial A_x}{\partial t}\right)+Q\left[\dot{y}\left(\frac{\partial A_y}{\partial x}-\frac{\partial A_x}{\partial y}\right)-\dot{z}\left(\frac{\partial A_x}{\partial z}-\frac{\partial A_z}{\partial x}\right)\right]$$

Corresponding equations are obtained for the time derivatives of the other cartesian components $(p_y - QA_y)$ and $(p_z - QA_z)$. Combining all three components yields:

$$\frac{d}{dt}(\mathbf{p}-Q\mathbf{A}) = -Q\left(\nabla\varphi+\frac{\partial\mathbf{A}}{\partial t}\right)+Q[\mathbf{v}\times(\nabla\times\mathbf{A})] \tag{18}$$

If we now substitute $m\mathbf{v}$ for $\mathbf{p}-Q\mathbf{A}$ according to (17) then eqn. (18) is seen to be the equation of motion (16) for charge carriers in an electromagnetic field. Therefore eqn. (16) is the correct Hamiltonian of a system consisting of a charge carrier in an electromagnetic field.

1.4. The Schrödinger equation

For a sufficiently detailed treatment of stimulated emission or absorption we will have to resort to the methods of quantum mechanics. Physical systems exhibiting stimulated emission or absorption change their states because of the interaction with electromagnetic fields. The time dependence of states in physical systems are quite generally described by Schrödinger's equation. The equation may be written in operator notation as

$$\bar{H}(\bar{p}_i, \bar{q}_i, t)\phi(q_i, t) = j\hbar\frac{\partial}{\partial t}\phi(q_i, t) \tag{19}$$

($j = \sqrt{(-1)}$; operator are indicator by a bar over the symbol).

Here $\phi\ (q_i, t)$ is the state function. The square of its absolute value $|\phi|^2$ is a measure of the probability that the system assumes coordinates q_i at time t. The normalization

$$\int \phi\phi^*\, d\tau = 1 \tag{20}$$

where the integration extends over the total range of each position coordinate q_i ensures that $|\phi|^2$ is the probability function for the state ϕ.

The coefficient $\hbar$ in (19) is, from the relation $\hbar = h/2\pi$, given by Planck's constant

$$h = 0{\cdot}6624\times10^{-33}\ \mathrm{VA}s^2$$

$\bar{H}$ is the "so-called" Hamiltonian operator. This operator is obtained from the classical Hamiltonian function for the particular system when both the p_i and the q_i are replaced by operators $\bar{p}_i$ and $\bar{q}_i$. The operators $\bar{p}_i$ and $\bar{q}_i$ for momentum and position in turn follow from their commutator $[\bar{q}, \bar{p}] = \bar{q}\bar{p}-\bar{p}\bar{q}$ which, according to one of the fundamental postulates of quantum mechanics, must satisfy:

$$[\bar{q}, \bar{p}] = j\hbar \tag{21}$$

If, as in Schrödinger's equation, we have chosen a representation ϕ (q_i, t) using the coordinates of position, then the operator $\bar{q}$ in eqn. (21) must also be replaced by the coordinates of position. Under these circumstances eqn. (21) is satisfied by a momentum operator, which has the following components

$$\bar{p}_i = -j\hbar \frac{\partial}{\partial q_i} \equiv -j\hbar \nabla_i \tag{22}$$

This is readily verified by substituting from (22) for $\bar{p}_i$ into (21).

In classical mechanics we have $\hbar = 0$. In this case $[\bar{q}_i, \bar{p}_i] = 0$ and the corresponding coordinates of position and momentum are said to commute. On the other hand in quantum mechanics $\hbar$ is finite and from (21) either the coordinates of position or of momentum are quantized. To this end, depending on the representation, either the position or the momentum coordinates are replaced by operators. The same operators also serve to quantize the Hamiltonian function and hence give the Hamiltonian operator. With the presentation of (19) in terms of the coordinates of position the Hamiltonian function is quantized to give the Hamiltonian operator when all components of momentum are replaced by differential operators according to (22).

If, for example, the Hamiltonian function of a particular system contains the x-component of momentum in cartesian coordinates $p_i = p_x = m\dot{x}$, then for the Hamiltonian operator it must be replaced by $-j\hbar\, \partial/\partial x$. If as in eqn. (16) the Hamiltonian function depends on the momentum vector $\mathbf{p}$, then the operator has to be replaced by the gradient $-j\hbar\nabla$ which in cartesian coordinates reads

$$\bar{\mathbf{p}} = -j\hbar \left(\mathbf{u}_x \frac{\partial}{\partial x} + \mathbf{u}_y \frac{\partial}{\partial y} + \mathbf{u}_z \frac{\partial}{\partial z} \right)$$

Here $\mathbf{u}_x$, $\mathbf{u}_y$, $\mathbf{u}_z$ are unit vectors in x-, y- and z-direction, respectively.

Schrödinger's equation (19) and the commutator condition (21) for the operators of position and momentum are introduced here as postulates without explaining them any further. They are assumed here to be laws which are valid quite generally and are justified sufficiently by experiments. This procedure is not unlike the usual procedure in electrical engineering where we start from Maxwell's equations. Corresponding to Maxwell's equations which are stated to describe completely all relevant phenomena in electrodynamics, Schrödinger's equation (19) and the commutator condition (21) are postulated in quantum mechanics.

Normally the Hamiltonian operator $\bar{H}$ can be separated into a stationary component $\bar{H}_s(\bar{p}_i, \bar{q}_i)$ which is independent of time and a dynamic component $\bar{H}_t(\bar{p}_i, \bar{q}_i, t)$ which changes with time:

$$\bar{H} = \bar{H}_s + \bar{H}_t$$

Let us assume first that the Hamiltonian operator is independent of time and is therefore the stationary component $\bar{H}_s$. Hamilton's function (16) is, for example, independent of time if there is no electromagnetic field or if all fields are static. Under these circumstances Schrödinger's equation may be separated by substituting for the state function the product

$$\phi(q_i, t) = \psi(q_i)X(t) \tag{23}$$

where ψ is a function of position and X a function of time only. From

$$X(t)\bar{H}_s\psi(q_i) = j\hbar\psi(q_i)\frac{dX(t)}{dt}$$

follows

$$\frac{\bar{H}_s\psi}{\psi} = j\frac{\hbar}{X}\frac{dX}{dt}$$

The left-hand side of this equation depends only on the coordinates q_i while the right-hand side is a function of time only. In order to satisfy this equation in all positions at all times each side has to be equal to a common constant quantity W. We therefore must have

$$H_s\psi = W\psi \tag{24}$$

and also

$$jh\frac{dX}{dt} = WX \tag{25}$$

The first of these two equations together with the boundary conditions for ψ, which must always be satisfied in physical systems, constitute an eigenvalue problem. There is always an infinite number of solutions with eigenfunctions ψ_n and eigenvalues W_n. Depending on the nature of the boundary conditions they form either a discrete spectrum or a continuous spectrum or a combination of discrete and continuous spectra.

Our interest here lies especially in a discrete spectrum of solutions ψ_n with eigenvalues W_n. These discrete solutions represent different states in which the physical system under consideration will remain stationary. W_n is always a real quantity and it gives the total energy of the system while it is in state ψ_n. The time dependence of the state function follows from the solution of the second equation

$$X_n(t) = e^{-j\frac{W_n}{\hbar}t}$$

If, for reasons which will be discussed in detail later, the state of the system changes from ψ_n to ψ_m, the total energy of the system also changes from W_n to W_m. The energy difference $W_m - W_n$, if positive, has to be supplied to the system. If it is negative it must be subtracted from it. If the change of state or transition is a radiating transition or a radiation absorbing transition a photon of energy $W_m - W_n$ and frequency $\omega = | W_m - W_n |/\hbar$ will either be emitted or absorbed.

According to (23) a particular solution of Schrödinger's equation for stationary states is

$$\phi_n(q_i, t) = \psi_n(q_i)e^{-j\frac{W_n}{\hbar}t}$$

The general solution is a combination of all these particular solutions

$$\phi(q_i, t) = \sum_n a_n\psi_n(q_i)e^{-j\frac{W_n}{\hbar}t} \tag{26}$$

The discrete state functions $\psi_n(q_i)$ form an orthogonal set:

$$\int \psi_n\psi_m^* \, d\tau = 0 \quad \text{if} \quad n \neq m$$

If they are normalized according to

$$\int \psi_n\psi_m^* \, d\tau = \delta_{nm}, \tag{27}$$

where

$$\delta_{nm}\begin{cases} = 1 & \text{if} \quad n = m \\ = 0 & \text{if} \quad n \neq m \end{cases}$$

is the Kronecker symbol, then from eqns. (20) and (27)

$$\sum_n | a_n |^2 = 1$$

Under these conditions $|a_n|^2$ is the probability that the physical system is in state n. The coefficients are constant and, in particular, are independent of time. Equation (26) therefore describes a general stationary state of the physical system.

1.5. Non-stationary solutions of Schrödinger's equation

The stationary solutions of Schrödinger's equation are obtained when its Hamilton operator is independent of time. They represent states in which the physical system remains as long as it is not interacting with external forces which are also changing with time. In the case of time-dependent influences a dynamic component $H_t(p_i, q_i, t)$ is added to the stationary component of the Hamilton operator. Schrödinger's equation must then be written as

$$(\overline{H}_s+\overline{H}_t)\phi(q_i, t) = j\hbar \frac{\partial\phi(q_i, t)}{\partial t} \tag{28}$$

To find a general solution we substitute

$$\phi(q_i, t) = \sum_n a_n(t)\psi_n(q_i)e^{-j\frac{W_n}{\hbar}t}$$

This trial function is similar to the solution of the stationary Schrödinger equation except that the coefficients are not now constant but are as yet unknown functions of time. Substituting into (28) we obtain:

$$\sum_n a_n\overline{H}_s\psi_n e^{-j\frac{W_n}{\hbar}t} + \sum_n \overline{H}_t a_n\psi_n e^{-j\frac{W_n}{\hbar}t} = j\hbar\left\{\sum_n \dot{a}_n\psi_n e^{-j\frac{W_n}{\hbar}} - \sum_n j\frac{W_n}{\hbar}a_n\psi_n e^{-j\frac{W_n}{\hbar}t}\right\} \tag{29}$$

Because the quantities ψ_n are solutions of the stationary Schrödinger equation, we have

$$\overline{H}_s\psi_n = W_n\psi_n.$$

The first sum on the left-hand side of (29) and the last sum on its right-hand side therefore cancel each other. Furthermore since $\overline{H}_t$ does not involve any differentiation with respect to time we have

$$\overline{H}_t a_n = a_n\overline{H}_t$$

Therefore eqn. (29) reduces to

$$\sum_n (j\hbar\dot{a}_n\psi_n - a_n\overline{H}_t\psi_n)e^{-j\frac{W_n}{\hbar}t} = 0$$

Let us multiply this equation by

$$\phi_m^* = \psi_m^* e^{j\frac{W_m}{\hbar}}$$

and integrate over the whole range of all coordinates q_i. Because of the orthonormality (27) of all ψ_n we then arrive at the following set of equations

$$j\hbar\dot{a}_m = \sum_n H_{mn}^{(t)} a_n e^{\frac{W_m - W_n}{\hbar}t} \qquad (m = 1, 2, \ldots) \tag{30}$$

Here we have written

$$H_{mn}^{(t)} = \int \psi_m^* \overline{H}_t \psi_n \, d\tau \tag{31}$$

This quantity is called the matrix element (m, n) of the operator $\overline{H}_t$ in the system of state functions ψ_n. Equation (31) defines the complete set of coefficients of a square matrix for $\overline{H}_t$.

With eqn. (30) the time-dependent Schrödinger equation has been transformed into a simultaneous set of ordinary first-order differential equations. The coefficients of this set of equations are functions of time as specified by the particular structure of the operator $\overline{H}_t(p_i, q_i, t)$. To solve this set of equations we note that for most arrangements in quantum electronics the effect of $\overline{H}_t$ is normally small compared with the effect of $\overline{H}_s$. The coefficients a_n will, therefore, vary only slowly with time, i.e. changes will take place only very gradually. Under these circumstances the set of eqns. (30) may, for given initial conditions, be solved approximately by repeated integration.

At time $t = 0$ let all the coefficients be specified as $a_n = a_n(0)$. This means that at $t = 0$ the physical system has the probability $|a_n(0)|^2$ of being in a state ψ_n. This probability distribution is now considered to be the zero order approximation to the solution of the differential equations (30). The $a_n(0)$ are substituted for all a_n on the left-hand sides. Integrating now over time we obtain the first order approximation to its solution

$$a_m(t) = a_m(0) - \frac{j}{\hbar}\sum_n \int_0^t H_{mn}^{(t)} a_n(0) e^{j\frac{W_m - W_n}{\hbar}t'} \, dt' \tag{32}$$

For the second approximation these functions $a_m(t)$ are in turn substituted for all a_n in the right-hand sides of (30) and another integration over time is performed. By successively repeating this integration the method converges to the actual solution of (30) under quite general assumptions. For most applications in quantum electronics the first order approximation (32) to the solution is sufficient.

The solution $a_m(t)$ describes the change of state of a physical system when external forces are acting upon it. More specifically the time functions $|a_m(t)|^2$ represent the probabilities that a physical system is in the corresponding state and tells us how this probability changes with time due to external forces.

The special initial condition

$$a_n(0) = \delta_{nm}$$

describes a system which is definitely in state m with state function ψ_m at time $t = 0$. Under this condition the first-order approximation (32) to the non-stationary solution is

$$a_m(t) = 1 - \frac{j}{\hbar}\int_0^t H_{mm}^{(t)}\, dt'$$

$$a_n(t) = -\frac{j}{\hbar}\int_0^t H_{nm}^{(t)} e^{j\frac{W_n - W_m}{\hbar}t'}\, dt'$$

In this case the physical meaning of the quantities $|a_m(t)|^2$ and $|a_n(t)|^2$ is quite clear:

$$w_{mm}(t) = |a_m(t)|^2 = \left|1 - \frac{j}{\hbar}\int_0^t H_{mm}^{(t')}\, dt'\right|^2$$

is the probability for the system at time t still to be in state m and

$$w_{nm}(t) = |a_n(t)|^2 = \frac{1}{\hbar^2}\left|\int_0^t H_{nm}^{(t')} e^{j\frac{W_n - W_m}{\hbar}t'}\, dt'\right|^2 \tag{33}$$

is the probability that the physical system during time t has made a transition to state n. Using these transition probabilities $w_{nm}(t)$ the processes in stimulated emission and absorption will be analysed.

1.6. The Hamiltonian operator of charge carriers in an electromagnetic field

To evaluate the approximate solution of the time dependent Schrödinger equation in the case of the interaction of matter with electromagnetic fields we have to know the Hamiltonian operator for such physical systems. To this end the Hamiltonian function is transformed into the corresponding Hamiltonian operator, a process which is also called quantization of the Hamiltonian function. In this quantization, according to the fundamental postulates of quantum mechanics, all momentum functions must be replaced by their respective momentum operators. We shall consider here a charge carrier in an electromagnetic field. Its Hamiltonian function (16) contains a momentum vector **p** which must be replaced by the differential operator $-j\hbar\nabla$. We obtain

$$\overline{H} = -\frac{\hbar^2}{2m}\nabla^2 + j\frac{\hbar Q}{2m}(\mathbf{A}\cdot\nabla+\nabla\cdot\mathbf{A}) + \frac{Q^2}{2m}A^2 + Q\varphi$$

According to the product rule for differentiation we have for any function $f(x)$ in the operator

$$\frac{\partial}{\partial x}f = \frac{\partial f}{\partial x} + f\frac{\partial}{\partial x}$$

Therefore we may write in Cartesian component form

$$A_x\nabla_x + \nabla_x A_x = 2A_x\nabla_x + \frac{\partial A_x}{\partial x}$$

and the corresponding vector operator is

$$\mathbf{A}\nabla + \nabla\mathbf{A} = 2\mathbf{A}\cdot\nabla + (\nabla\cdot\mathbf{A}).$$

Also in the Hamiltonian operator

$$\overline{H} = -\frac{\hbar^2}{2m}\nabla^2 + j\frac{hQ}{m}\mathbf{A}\cdot\nabla + j\frac{\hbar Q}{2m}(\nabla\cdot\mathbf{A}) + \frac{Q^2}{2m}A^2 + Q\varphi$$

the term with $(\nabla\cdot\mathbf{A})$ is now merely a multiplier.

The Hamiltonian operator will now be separated into a stationary component $\overline{H}_s$ which is independent of time and a dynamic component $\overline{H}_t$, which changes with time. To this end the electromagnetic potentials must first be separated into static and dynamic components

$$\mathbf{A} = \mathbf{A}_s + \mathbf{A}_t \qquad \varphi = \varphi_s + \varphi_t$$

We will assume here that $\varphi_t = 0$, because in source free regions any dynamic field may be represented by $\mathbf{A}_t$ alone, where, in addition $\nabla \cdot \mathbf{A}_t = 0$. Using the position vector $\mathbf{r}$ referred to an arbitrary fixed point, the static fields are derived from

$$\varphi_s = \mathbf{r} \cdot \mathbf{E}_s \quad \text{and} \quad \mathbf{A}_s = \tfrac{1}{2}(\mathbf{B}_s \times \mathbf{r})$$

Separated into static and dynamic components the Hamiltonian operator is

$$\overline{H}_s = -\frac{\hbar^2}{2m}\nabla^2 + U + j\frac{\hbar Q}{m}\mathbf{A}_s \cdot \nabla + \frac{Q^2}{2m}A_s^2 + Q\varphi_s$$

$$\overline{H}_t = j\frac{\hbar Q}{m}\mathbf{A}_t\left(\nabla - j\frac{Q}{\hbar}\mathbf{A}_s\right) + \frac{Q^2}{2m}A_t^2 \tag{34}$$

Here $\nabla \cdot \mathbf{A}_s = 0$ has been taken into account.

The stationary Hamiltonian operator contains a function U which is meant to be a potential energy not included in $\mathbf{A}_s$ or φ_s. U may, for example, account for the potential of the atomic or molecular system to which the charge carrier is bound by Coulomb forces.

In many situations in quantum electronics no external static fields are applied. Also in many cases such external static fields do not affect induced transitions. We will therefore omit φ_s and A_s in the operators and consider only the simpler forms:

$$\overline{H}_s = -\frac{\hbar^2}{2m}\nabla^2 + U$$

$$\overline{H}_t = j\frac{\hbar Q}{m}\mathbf{A} \cdot \nabla + \frac{Q^2}{2m}A^2 \tag{35}$$

Also the subscript t of $\mathbf{A}_t$ has been dropped for convenience.

We must note, however, that a specific elementary system, such as for example an atom or a molecule has not just one electron but a number of them, say N. All these N electrons are bound by a Coulomb potential U to the nucleus. Under these circumstances the state functions ϕ_n or ψ_n depend on all of the $3N$ position coordinates of the electrons. The differential operation ∇ must then be applied with respect to all $3N$ coordinates. For the ith electron ∇_i in Cartesian coordinates is

$$\nabla_i = \mathbf{u}_x \frac{\partial}{\partial x_i} + \mathbf{u}_y \frac{\partial}{\partial y_i} + \mathbf{u}_z \frac{\partial}{\partial z_i}$$

For a multi-electron system with the elementary electronic charge $Q = -e$ we have, therefore,

$$\overline{H}_s = -\frac{\hbar^2}{2m}\sum_i \nabla_i^2 + U$$

$$\overline{H}_t = -j\frac{e\hbar}{m}\sum_i A_i\nabla_i + \frac{e^2}{2m}\sum_i A_i^2 \tag{36}$$

The vector $\mathbf{A}_i$ in the latter operator must be evaluated at the location $\mathbf{r}_i$ of the ith electron.

1.7. Approximations for the interaction operator

The method which we are utilizing here to analyse the interaction between electromagnetic fields and matter is not entirely a quantum mechanical approach. It is true that we have quantized the states of atoms and molecules by representing them as solutions of Schrödinger's equation. We have not however quantized the electromagnetic field which is interacting with them. In order to close this gap we simply assume that the change in total energy $W_m - W_n$ when there is a transition from one state m to another n will be subtracted from or added to the electromagnetic field energy in form of a photon of energy

$$\hbar\omega = |W_m - W_n|$$

This assumption is justified by experimental evidence. Furthermore experiments even show that each photon of induced emission is added with the same polarization and in phase with the electromagnetic oscillation or wave which has stimulated the transition. From a fully quantum mechanical treatment these assumptions and the supporting experimental evidence could be directly deduced.

In our present semi-classical treatment it will not be necessary to take the Hamiltonian operator for interaction between charge carriers and field into account exactly. When considering only weak interactions the term which contains the square of the vector potential may be neglected in $\overline{H}_t$. The assumptions made for the semi-classical treatment neglect effects which would be of the same order as this squared term. Taking this term into account would, therefore, not improve our calculation which is only of an approximate nature anyway. Thus from here on we will let

$$\bar{H}_t = -j\frac{e\hbar}{m}\sum_{i=1}^{N} \mathbf{A}_i \nabla_i$$

This operator will now be developed further and evaluated for more specific field distributions.

First let us calculate the operator for *electric dipole interaction.* In this case we assume a field distribution which is uniform in space over the extent of the elementary microsystem. The vector potential $\mathbf{A}_i = \mathbf{A}(\mathbf{r}_i, t)$ may be replaced by $\mathbf{A}_0 = \mathbf{A}(0, t)$ at all positions $\mathbf{r}_i$ of each electron where the reference point $\mathbf{r} = 0$ has been chosen to be the centre of the micro-system. Normally it should be the position of the nucleus of the atom or centre of gravity in the case of a molecule. Thus evaluating the $\mathbf{A}_i$ we obtain

$$\bar{H}_t = -j\frac{e\hbar}{m} A_0 \sum_i \nabla_{iA} \tag{37}$$

where ∇_{iA} is the component of the gradient ∇_i in the direction of $\mathbf{A}_0$.

Transitions between states and the corresponding transition probabilities are determined by the matrix elements of the interaction operator. They are in the case of $\bar{H}_t$ from (37)

$$H_{mn}^{(t)} = -j\frac{e\hbar}{m} A_0 \sum_i \int \psi_m^* \nabla_{iA} \psi_n \, d\tau \tag{38}$$

In order to find a physical interpretation for these matrix elements we will now transform them to more suitable expressions. From the stationary Schrödinger equation

$$\bar{H}_s \psi_n = W_n \psi_n \quad \text{resp.} \quad \bar{H}_s^* \psi_m^* = W_m \psi_m^*$$

it follows that

$$\int \psi_m^* \mathbf{r}_i \bar{H}_s \psi_n \, d\tau = W_n \int \psi_m^* \mathbf{r}_i \psi_n \, d\tau \tag{39}$$

as well as

$$\int (\bar{H}_s^* \psi_m^*) \mathbf{r}_i \psi_n \, d\tau = W_m \int \psi_m^* \mathbf{r}_i \psi_n \, d\tau \tag{40}$$

We recall here that the eigenvalues W_n and W_m are the total energies of the respective states and always real quantities. The Hamilton operator is Hermitian which means that

$$\int (\overline{H}_s^* \psi_m^*) \mathbf{r}_i \psi_n \, d\tau = \int \psi_m^* \overline{H}_s \mathbf{r}_i \psi_n \, d\tau$$

Subtracting eqn. (39) from eqn. (40) and using the Hermitian character of $\overline{H}_s$ yields

$$\int \psi_m^* (\overline{H}_s \mathbf{r}_i - \mathbf{r}_i \overline{H}_s) \psi_n \, d\tau = (W_m - W_n) \int \psi_m^* \mathbf{r}_i \psi_n d\tau \tag{41}$$

Furthermore we note that for the operator

$$\overline{H}_s = -\frac{\hbar^2}{2m} \sum_i \nabla_i^2 + U$$

we have

$$(U\mathbf{r}_i - \mathbf{r}_i U) \equiv 0$$

as well as

$$(\nabla_k^2 \mathbf{r}_i - \mathbf{r}_i \nabla_k^2) \equiv 0 \quad \text{for} \quad i \neq k$$

since the operators U and ∇_k commute with $\mathbf{r}_i$, provided $i \neq k$. Equation (41) therefore reduces to

$$\frac{\hbar^2}{2m} \int \psi_m^* (\nabla_i^2 \mathbf{r}_i - \mathbf{r}_i \nabla_i^2) \psi_n \, d\tau = (W_m - W_n) \int \psi_m^* \mathbf{r}_i \psi_n \, d\tau$$

Noting in addition that

$$\nabla_i^2 (\mathbf{r}_i \psi_n) \equiv r_i \nabla_i^2 \psi_n + 2 \nabla_i \psi_n$$

we finally obtain

$$\int \psi_m^* \nabla_i \psi_n \, d\tau = -\frac{m}{\hbar^2} (W_m - W_n) \int \psi_m^* \mathbf{r}_i \psi_n \, d\tau$$

This is just the integral whose components appear in the expression (38) for the matrix elements. We may therefore substitute for these components in (38) from the present vector integral. The result is

$$H_{mn}^{(t)} = -j \frac{W_m - W_n}{\hbar} A_0 (\mathbf{P}_{mn})_A \tag{42}$$

where

$$(\mathbf{P}_{mn})_A = \int \psi_m^* P_A \psi_n \, d\tau \tag{43}$$

is the matrix element (m, n) of the component P_A of the classical dipole moment

$$\mathbf{P} = -\sum_i e\mathbf{r}_i$$

P_A is the component of vector $\mathbf{P}$ in the direction of the vector potential $\mathbf{A}_0$ and hence in the direction of the external electric field vector

$$\mathbf{E} = -\frac{\partial \mathbf{A}}{\partial t}$$

in the case of linear polarization. In this first approximation the interaction with the electromagnetic field $\mathbf{A}$ is only through the electric field vector $\mathbf{E}$ which acts upon the electric dipole moment $\mathbf{P}$ of the elementary system. Therefore this first order approximation is called the *electric dipole interaction.*

The electric dipole interaction is only an approximation since the electromagnetic field has been assumed uniform in the region of space over which the elementary system extends. To improve upon this approximation and possibly to obtain also the interaction with the magnetic field vector we now take the gradients of the electromagnetic field, at least to the first order, into account. To this end the interaction operator (37) will be supplemented by the following term

$$\overline{H}_t' = -j\frac{e\hbar}{m}\sum_i \mathbf{r}_i(\nabla\mathbf{A})_0\cdot\nabla_i$$

Here the subscript 0 of $(\nabla\mathbf{A})_0$ indicates that the gradients of the vector components of $\mathbf{A}$ are to be evaluated at $\mathbf{r} = 0$ when substituting into the above expression. This means that in a Taylor expansion around the reference point $\mathbf{r} = 0$ only terms up to first order are to be considered. In Cartesian coordinates we have

$$\mathbf{r}_i(\nabla\mathbf{A})_0 = \mathbf{u}_x\left(x_i\frac{\partial A_x}{\partial x}+y_i\frac{\partial A_x}{\partial y}+z_i\frac{\partial A_x}{\partial z}\right)+\mathbf{u}_y\left(x_i\frac{\partial A_y}{\partial x}+y_i\frac{\partial A_y}{\partial y}+z_i\frac{\partial A_y}{\partial z}\right)$$
$$+\mathbf{u}_z\left(x_i\frac{\partial A_z}{\partial x}+y_i\frac{\partial A_z}{\partial y}+z_i\frac{\partial A_z}{\partial z}\right)$$

and

$$\mathbf{r}_i(\nabla\mathbf{A})_0\nabla_i = \left(x_i\frac{\partial A_x}{\partial x}+y_i\frac{\partial A_x}{\partial y}+z_i\frac{\partial A_x}{\partial z}\right)\frac{\partial}{\partial x_i}$$
$$+\left(x_i\frac{\partial A_y}{\partial x}+y_i\frac{\partial A_y}{\partial y}+z_i\frac{\partial A_y}{\partial z}\right)\frac{\partial}{\partial y_i}+\left(x_i\frac{\partial A_z}{\partial x}+y_i\frac{\partial A_z}{\partial y}+z_i\frac{\partial A_z}{\partial z}\right)\frac{\partial}{\partial z_i}$$

If the gradients of all the vector components are expressed in the form

$$\frac{\partial A_x}{\partial y} = \frac{1}{2}\left(\frac{\partial A_x}{\partial y} - \frac{\partial A_y}{\partial x}\right) + \frac{1}{2}\left(\frac{\partial A_x}{\partial y} + \frac{\partial A_y}{\partial x}\right)$$

then the magnetic induction $\mathbf{B} = \nabla \times \mathbf{A}$ may be introduced

$$\frac{\partial A_x}{\partial y} = -\frac{1}{2} B_z + B_{xy}$$

Here, in addition, we have written

$$B_{xy} = \frac{1}{2}\left(\frac{\partial A_x}{\partial y} + \frac{\partial A_y}{\partial x}\right)$$

for convenience. Similarly we obtain

$$\frac{\partial A_y}{\partial x} = \frac{1}{2} B_z + B_{yx}$$

The additional term $\overline{H}_t'$ of the interaction operator may now be separated into

$$\overline{H}_t' = \overline{H}_{tM}' + \overline{H}_{tQ}' \tag{44}$$

where

$$\overline{H}_{tM}' = -j\frac{e\hbar}{2m}\sum_i \left\{ B_x\left(y_i\frac{\partial}{\partial z_i} - z_i\frac{\partial}{\partial y_i}\right) + B_y\left(z_i\frac{\partial}{\partial x_i} - x_i\frac{\partial}{\partial z_i}\right) + B_z\left(x_i\frac{\partial}{\partial y_i} - y_i\frac{\partial}{\partial x_i}\right)\right\}$$

and

$$\overline{H}_{tQ}' = -j\frac{e\hbar}{m}\sum_i \left\{ B_{xx}x_i\frac{\partial}{\partial x_i} + B_{yy}y_i\frac{\partial}{\partial y} + B_{zz}z_i\frac{\partial}{\partial y_i} + B_{xy}\left(x_i\frac{\partial}{\partial y_i} + y_i\frac{\partial}{\partial x_i}\right) + B_{yz}\left(y_i\frac{\partial}{\partial z_i} + z_i\frac{\partial}{\partial y_i}\right) + B_{xz}\left(x_i\frac{\partial}{\partial z_i} + z_i\frac{\partial}{\partial x_i}\right)\right\}$$

The interaction which is described by the operator component $\overline{H}_{tM}'$ is called the *magnetic dipole interaction*. Through **B** the electromagnetic oscillation interacts directly by its magnetic field. $\overline{H}_{tQ}'$ describes the "so-called" electric quadrupole interaction. $\overline{H}_{tQ}'$ will not be considered any further.

The magnetic dipole interaction can be represented more directly in terms of the angular momentum

$$\mathbf{D} = \sum_i \mathbf{r}_i \times \mathbf{p}_i$$

of the particular elementary system. In order to quantize **D** and obtain the angular momentum operator $\bar{\mathbf{D}}$ from the angular momentum vector all linear momentum vectors **p** must again be replaced by their vector operators according to

$$\bar{\mathbf{p}}_i = -j\hbar\nabla_i$$

for the ith electron in a multi-electron system. The angular momentum operator is thus written as

$$\bar{\mathbf{D}} = -j\hbar \sum_i \mathbf{r}_i \times \nabla_i \tag{45}$$

The operator $\bar{\mathbf{D}}$ has the following Cartesian components

$$\bar{D}_x = -j\hbar \sum_i \left(y_i \frac{\partial}{\partial z_i} - z_i \frac{\partial}{\partial y_i} \right); \qquad \bar{D}_y = -j\hbar \sum_i \left(z_i \frac{\partial}{\partial x_i} - x_i \frac{\partial}{\partial z_i} \right);$$

$$\bar{D}_z = -j\hbar \sum_i \left(x_i \frac{\partial}{\partial y_i} - y_i \frac{\partial}{\partial x_i} \right)$$

Comparing these components of the angular momentum operator with the cartesian components of $\bar{H}'_{tM}$ it is seen that the latter may be written as

$$\bar{H}'_{tM} = \frac{e}{2m} \mathbf{B} \cdot \bar{\mathbf{D}}$$

Here in analogy to the classical magnetic dipole moment

$$\mathbf{M} = -\frac{e}{2m} \mathbf{D}$$

an operator for the magnetic dipole moment can be defined as

$$\bar{\mathbf{M}} = -\frac{e}{2m} \bar{\mathbf{D}}$$

Finally the Hamiltonian operator for magnetic dipole interaction in terms of this new dipole operator is

$$\bar{H}'_{tM} = -\mathbf{B} \cdot \bar{\mathbf{M}} \tag{46}$$

This form of the operator corresponds entirely to the operator for the electric dipole interaction.

1.8. Interaction with sine waves

Up to now the discussion and resulting formulae have been quite general in that electromagnetic fields of arbitrary time dependence were considered. We will now assume, however, that all field components are purely sinusoidal functions of time. This does not correspond to any actual situation because hardly any process is purely sinusoidal in time. But any arbitrary time dependence can always be represented by its spectral components which, indeed, have this idealized time dependence. Also a corresponding specialization is adequate for the behaviour of the electromagnetic field with respect to the spatial coordinates. It is sufficient to assume a uniform plane wave. Any arbitrary field distribution may be expressed in terms of uniform plane waves of different polarizations and directions of propagation.

We choose a uniform plane wave whose vector potential is

$$\mathbf{A} = \mathbf{u}_y \sqrt{2}\frac{E_0}{\omega}\cos(\omega t - kz) = \mathbf{u}_y \frac{E_0}{\sqrt{2}\omega}\left[e^{j(\omega t - kz)} + e^{-j(\omega t - kz)}\right] \qquad (47)$$

This wave travels in the positive z-direction and is linearly polarized in the $(y-z)$ plane. The electric field vector is

$$\mathbf{E} = \mathbf{u}_y \sqrt{2} E_0 \sin(\omega t - kz)$$

Let this field interact with an elementary system whose centre is located at $z = 0$ and which at time $t = 0$ is in state m with the state function ψ_m. The probability (33) that up to time t the system has undergone a transition to another state n due to this interaction is, under these circumstances,

$$w_{nm}(t) = \frac{\omega_{nm}^2}{2\hbar^2\omega^2} E_0^2 |(\mathbf{P}_{nm})_y|^2 \left| \frac{e^{j(\omega+\omega_{nm})t} - 1}{\omega + \omega_{nm}} - \frac{e^{-j(\omega-\omega_{nm})t} - 1}{\omega - \omega_{nm}} \right|^2 \qquad (48)$$

Here we have used the notation

$$\omega_{nm} = \frac{W_m - W_m}{\hbar} \qquad (49)$$

because this quantity not only has the dimension of frequency but also appears in (48) together with the angular frequency ω of the external a.c. field. ω_{nm} is called the transition frequency between states m and n.

Equation (48) again contains the matrix element (n, m) of the y-component of the electric dipole moment. Since the matrix element of any such component is defined by eqn. (45), the expression (48) may still be generalized somewhat. The matrix element in (48) is always based on the component of the dipole moment in the direction of the external electric field. All that counts in (48) is, therefore, the scalar product of both these vectors

$$w_{nm}(t) = \frac{1}{2\hbar^2}\frac{\omega_{nm}^2}{\omega^2}|\mathbf{E}_0\cdot\mathbf{P}_{nm}|^2\left|\frac{e^{j(\omega+\omega_{nm})t}-1}{\omega+\omega_{nm}}-\frac{e^{-j(\omega-\omega_{nm})t}-1}{\omega-\omega_{nm}}\right|^2 \tag{50}$$

In this expression $\mathbf{E}_0$ is meant to be the vector which is formed by the complex phasors of the electric field components, while $\mathbf{P}_{nm}$ is a vector formed by the matrix elements of the vector components of the dipole moment $\mathbf{P}$.

For the transition probability in eqn. (50) both terms of the last squared factor are functions of time. But at most only one of these terms will contribute significantly to the transition probability and this only when the denominator of the particular term is very small. In other words a transition will have a significant probability only if either

$$\omega \simeq \omega_{nm} = \frac{W_m - W_n}{\hbar}$$

or

$$\omega \simeq -\omega_{nm} = \frac{W_m - W_n}{\hbar}$$

With one or the other of these conditions for the frequency of oscillation satisfied the smaller of the terms in eqn. (50) may be neglected. The result is

$$w_{nm}(t) = \frac{2}{\hbar^2}|\mathbf{E}_0\cdot\mathbf{P}_{nm}|^2\frac{\sin^2\frac{1}{2}(\omega-\omega_{nm})t}{(\omega-\omega_{nm})^2} \tag{51}$$

Figure 1 shows the dependence of this transition probability on frequency ω for different durations of the interaction between external fields and the elementary system.

If only a short time has elapsed since the field started to interact with the micro-system the transition probability is small, even when the external frequency ω is in resonance with the transition frequency ω_{nm}:

$$\Delta\omega \equiv \omega - |\omega_{nm}| = 0$$

Also if, for this short interaction time, the external frequency is changed from the transition frequency ($\Delta\omega \neq 0$), the transition probability decreases only gradually. If the interaction between the field and the micro-system lasts longer then the transition probability at resonance increases steadily. But for any change of the external frequency from resonance it decreases very rapidly. The resonance characteristic of the transition probability as a function of external frequency will be the more pronounced and sharper the longer the interaction lasts.

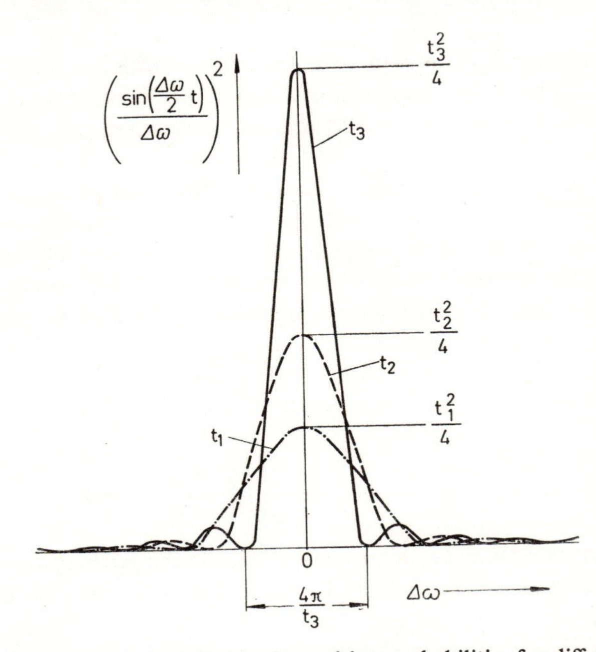

FIG. 1. Frequency characteristics of transition probabilities for different time intervals of interaction

Whether the micro-system makes a transition from state n to state m or from m to n depends on the initial state. Whether in this process energy is absorbed from the field or added to the field depends on the sign of

$$\Delta W_{mn} = W_m - W_n$$

If $\Delta W_{mn} > 0$ this energy is released from the micro-system and we have stimulated emission. On the other hand if $\Delta W_{mn} < 0$ the micro-system takes up this energy. In this case the field induces absorption.

As may be seen from eqn. (43) the following symmetry holds for the matrix elements

$$|\mathbf{P}_{nm}| = |\mathbf{P}_{mn}|$$

Consequently we have at all times $w_{nm}(t) = w_{mn}(t)$ as may be verified from the general expression (50). The probability of a micro-system which is initially in state m and which is interacting with an external field making a transition from state m to state n is just as large as the probability for a transition in the opposite direction, provided the micro-system is initially in the state n. In this sense stimulated emission and absorption are equally probable.

1.9. Interaction with thermal radiation

The uniform plane wave which is purely sinusoidal in time only approximates to practical radiation fields or is just a spectral component of such fields. Signals which, for example, are to be amplified, change their amplitude and phase with time. They consist of a frequency spectrum of finite width.

Before considering such actual signals we will analyse another special case, namely the interaction of a micro-system with thermal radiation. The field of such natural and random radiation may be represented by plane waves which are polarized and travelling in all directions of space and whose frequency components form a continuous spectrum. All these spectral components and polarizations are entirely independent from one another.

The direction of propagation is irrelevant for the electric dipole interaction, only the direction of polarization of the electric field counts. Furthermore, only the squares of the amplitudes of the electric field components enter the probability formulae for stimulated transitions. Therefore we will separate the field of thermal radiation into its three orthogonal directions of polarization and evaluate the square of the amplitude for each. For the amplitude of a single plane wave which is polarized in the x-direction we have

$$|E_x|^2 = \eta S_m$$

Here S_m is the real part of the complex Poynting vector, i.e. S_m is the mean value of the energy flow or radiation density. $\eta = \sqrt{\mu_0/\varepsilon_0}$ is the wave impedance of free space.

To take all spectral components of this polarization within the spectral range from say ω_1 to ω_2 into account, we only need to add the real parts of their complex Poynting vectors. Since these spectral components are uncorrelated in time all the cross terms will not contribute to the mean value, but will disappear when we average over time. For a continuous spectrum the sum of all spectral components becomes an integral with ω the variable of integration. The intensity of the x-polarization in the spectral range between ω_1 and ω_2 is therefore

$$I_x = \frac{1}{2\pi} \int_{\omega_1}^{\omega_2} I_x(\omega)\, d\omega$$

where $I_x(\omega)\, d\omega/2\pi = S_m = |E_x|^2/\eta$ is the mean value of energy flow in the spectral interval df at f. $I_x(\omega)$ is the intensity per cycle, also called spectral intensity. Considering the velocity of propagation $c = 1/\sqrt{\mu_0 \varepsilon_0}$ of all waves, the following spectral energy density is associated with the above spectral intensity

$$\sigma_x(\omega) = \frac{1}{c} I_x(\omega)$$

To be more specific $\sigma_x(\omega)$ is the electromagnetic energy per unit volume per cycle which is associated with the polarization E_x at ω. For thermal radiation the three Cartesian components $\sigma_x(\omega)$, $\sigma_y(\omega)$ and $\sigma_z(\omega)$ of the total spectral energy density are of equal magnitude. Therefore we have

$$\sigma(\omega) = \sigma_x(\omega) + \sigma_y(\omega) + \sigma_z(\omega)$$

where

$$\sigma_x(\omega) = \sigma_y(\omega) = \sigma_z(\omega) = \tfrac{1}{3}\sigma(\omega)$$

We will now have to evaluate the transition probability (51) for this spectral energy density. Using our present notation we obtain

$$\begin{aligned} |\mathbf{E}_0 \cdot \mathbf{P}_{nm}|^2 &= |E_x(\mathbf{P}_{nm})_x|^2 + |E_y(\mathbf{P}_{nm})_y|^2 + |E_z(\mathbf{P}_{nm})_z|^2 \\ &= \eta(I_x(\omega)|\mathbf{P}_{nm}|_x^2 + I_y(\omega)|\mathbf{P}_{nm}|_y^2 + I_z(\omega)|\mathbf{P}_{nm}|_z^2)\frac{d\omega}{2\pi} \\ &= \frac{1}{3\varepsilon_0}\sigma(\omega)(|\mathbf{P}_{nm}|_x^2 + |\mathbf{P}_{nm}|_y^2 + |\mathbf{P}_{nm}|_z^2)\frac{d\omega}{2\pi} \end{aligned} \tag{52}$$

So far we have taken into account all polarizations of the thermal radiation restricted to the spectral interval $d\omega$. To account for the interaction with all spectral components (51) must be integrated over the full spectral range. From eqn. (51) using eqn. (52) we obtain

$$w_{nm}(t) = \frac{1}{3\pi\varepsilon_0\hbar^2}(|\mathbf{P}_{nm}|_x^2+|\mathbf{P}_{nm}|_y^2+|\mathbf{P}_{nm}|_z^2)\int\limits_{-\infty}^{\infty}\sigma(\omega)\frac{\sin^2\frac{1}{2}(\omega-\omega_{nm})t}{(\omega-\omega_{nm})}d\omega \tag{53}$$

To evaluate the integral in this expression we will not make any use of the laws for spectral distribution of thermal radiation as yet. Rather we will assume that the interaction lasts long enough for the factor $\frac{1}{\Delta\omega^2}\sin^2\frac{1}{2}\Delta\omega t$ to have a very pronounced resonance peak at $\Delta\omega = 0$. The situation is shown in Fig. 1. Compared with this resonance peak $\sigma(\omega)$ changes only gradually with frequency. We may therefore let $\sigma(\omega) = \sigma(\omega_{nm})$ be a factor which is independent of the variable of integration in (53).

Using the definite integral

$$\int\limits_{-\infty}^{\infty}\frac{\sin^2\frac{1}{2}\Delta\omega t}{\Delta\omega^2}d\omega = \frac{\pi}{2}t$$

we thus obtain

$$w_{nm}(t) = \frac{1}{6\varepsilon_0\hbar^2}\sigma(\omega_{nm})(|\mathbf{P}_{nm}|_x^2+|\mathbf{P}_{nm}|_y^2+|\mathbf{P}_{nm}|_z^2)\,t$$

The interaction with thermal radiation causes transitions with probabilities which increase in direct proportion to the duration of the interaction. The transition probability is also proportional to the spectral energy density of the random radiation at the transition frequency.

These simple proportionality relations may be expressed by

$$w_{nm}(t) = B_{nm}\cdot\sigma(\omega_{nm})\cdot t \tag{54}$$

where the proportionality factor is

$$B_{nm} = \frac{1}{6\varepsilon_0\hbar^2}|P_{nm}|^2$$

and

$$|P_{nm}|^2 = |\mathbf{P}_{nm}|_x^2+|\mathbf{P}_{nm}|_y^2+|\mathbf{P}_{nm}|_z^2 \tag{55}$$

B_{nm} is called the *Einstein coefficient* of stimulated emission. It was first calculated by Einstein when he investigated this interaction with thermal radiation. Since

$$|P_{nm}|^2 = |P_{mn}|^2$$

we also have

$$B_{nm} = B_{mn} \tag{56}$$

The Einstein coefficient of stimulated emission is equal to the corresponding coefficient for absorption. Stimulated emission is as probable as absorption for any one particular micro-system.

So far we have always considered only one single atomic or molecular system. For such an elementary system we have calculated the probability of stimulated transitions due to external fields. As is customary these single atoms or molecules have occasionally been called micro-systems. The macroscopic physical arrangements do not consist of just one atomic or molecular micro-system. There is, rather, a very large number of them, always together. Such a collection of a large number of micro-systems is called a macro-system. However, it is usual in physical arrangements of quantum electronics for all micro-systems which participate in stimulated transitions to be nearly identical and nearly independent of one another. We are therefore justified in applying the results for a single micro-system directly to the corresponding macro-systems.

It will be assumed henceforth that the macro-system consists of a large number of identical micro-systems which are all independent of one another. Let the energy level diagram of each micro-system have two levels W_1 and W_2 as shown in Fig. 2. All other energy levels which are not shown in Fig. 2 will be disregarded here. For convenience we will assume that the two levels in Fig. 2 are not degenerate. The micro-systems have, therefore, only one state 1 described by its state-function ψ_1, which has an associated energy W_1 and, also, only one state 2 with state function ψ_2, and corresponding energy W_2. All other states of the micro-systems shall have different energies. Between state 1 and state 2 electric dipole transitions are assumed to be possible, i.e. all or at least one component of the matrix element $\mathbf{P}_{12}$ shall be non-zero. This electric dipole transition is then also called an allowed transition. All micro-systems of the macroscopic arrangement are assumed to be exposed to one and the same radiation field.

If N_2 of these micro-systems are in state 2 then according to the probability (54) the number of transitions to state 1 per unit time or, in other words, the transition rate, is

$$w'_{21} = B_{21}\sigma(\omega_{12})N_2 \tag{57}$$

Here also it has been assumed that the spectral energy density $\sigma(\omega)$ is independent of frequency in the spectral range near ω_{12} where there is significant interaction with the radiation field.

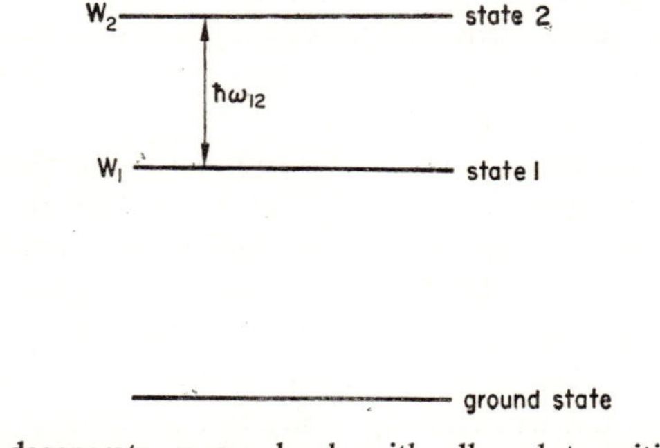

FIG. 2. Non-degenerate energy levels with allowed transitions in micro-systems

Under these conditions the transition rate w'_{21} is independent of time. Equation (57) describes a stationary state of the macro-system. There are of course other transitions, particularly those from state 1 to 2, so that the population N_2 of state 2 will remain constant in time.

In lasers and masers the interaction is normally with electromagnetic fields of high spectral purity. For the ideal case of a monochromatic wave of mean energy flow S_m and frequency ω' the spectral energy density may be represented by Dirac's delta function

$$\sigma(\omega) = \frac{S_m}{c}\,\delta(\omega' - \omega)$$

With this particular spectral distribution of energy density a transition probability for each single micro-system would be obtained from (53) which depends on frequency ω' as indicated by expression (51) and as shown in Fig. 1. For the macroscopic arrangement the number of transitions per unit time will then likewise depend on ω' and t. The rate of transitions will be largest for $\omega' = \omega_{12}$. The longer the interaction with a monochromatic wave of frequency $\omega' = \omega_{12}$ lasts the more transitions will be made. But if $\omega' \neq \omega_{12}$ the transition rate will fall off rapidly. When the interaction lasts long enough transitions will be made only when the external field is at resonance with ω_{12}. For even very slight detuning, w_{21} from eqn. (51) for the micro-system, and w'_{21} for the macro-system will be insignificant. In addition w'_{21} is dependent on time. Hence from these equations no stationary state will ever be established. This does not agree with experimental evidence. Obviously eqn. (51) does not hold for interactions of arbitrarily long duration.

In order to analyse the phenomena associated with long duration interactions with monochromatic waves we will first have to consider a different kind of transition between states that occur in micro-systems in addition to stimulated transitions.

These additional transitions will be found to limit the lifetime of a micro-system in a definite state, such as state 2, even if there is no external field. When the lifetime in a particular state is limited the time of interaction between a field and the micro-system in this particular state must be limited also. For limited interaction time transition probability and transition rate will have a frequency characteristic of finite band width. Let us therefore study this additional type of transition between states first.

1.10. Spontaneous transitions

Atomic oscillators not only emit energy when they are interacting with an external electromagnetic field. They also radiate when they are not exposed to any external forces. This independent emission of energy may also be observed for the classical model of an electron oscillating in a potential well. The explanation based on the classical model will again serve to clarify the quantum-mechanical treatment.

An electron which is oscillating according to eqn. (1) radiates power

$$P = \frac{e^2 r_0^2 \omega^4}{3c^3}$$

The total energy of the electronic oscillation is

$$W = \tfrac{1}{2} m\omega^2 r_0^2$$

Through radiation this power decreases exponentially with a time constant

$$\tau = \frac{W}{P} = 6\pi \frac{mc^3\varepsilon}{\omega^2 e^2} \tag{58}$$

The quantity τ is called the classical lifetime of the electron.

Quantum mechanically the electron can only remain stationary in certain states with constant energy within the Coulomb field of the remainder of the atom or molecule. There is no possibility for continuous emission of energy. The electron may however make a transition from a state of higher energy to a state of lower energy. The energy difference will then be emitted in form of a photon. Such a transition happens spontaneously without any external forces acting. Transitions of this type are even necessary just to maintain thermodynamic equilibrium under normal conditions. This will be seen as follows:

For a large number of micro-systems with energy levels as shown in Fig. 2 the spontaneous and induced transitions from state 2 to state 1 must balance the induced transitions from 1 to 2 in order to maintain equilibrium. We therefore have, for the three transition rates,

$$\sigma(\omega_{12})\, B_{21} N_1 = A_{12} N_2 + \sigma(\omega_{12})\, B_{12} N_2 \tag{59}$$

The rate for spontaneous transitions is $A_{12} \cdot N_2$; because, just as in case of induced transitions, the rate must be proportional to the number N_2 of micro-systems in state 2. The factor A_{12} is called the Einstein coefficient of spontaneous emission. In a micro-system the coefficient specifies the number of spontaneous transitions which are made on the average per unit time for each micro-system which is in state 2. For the individual micro-system in state 2 it therefore specifies the average number of spontaneous transitions from 2 to 1 per unit time. From this its inverse

$$\tau_{12} = \frac{1}{A_{12}}$$

is seen to be the average lifetime of a micro-system in state 2. This average lifetime has the same meaning as the lifetime of the previous classical electron oscillator. From quantum mechanical considerations we will however obtain a different form for τ_{12} than that obtained in (58) for the classical lifetime.

The average lifetime of individual micro-systems in the physical arrangement will be given correctly by τ_{12} only under certain conditions.

These conditions are: that no external forces are interacting which, by inducing transitions to 1, would shorten the lifetime in 2, and also that no induced or spontaneous transitions to other states but state 1 may be made. All additional transitions would further shorten the lifetime.

The coefficient A of the spontaneous emission may be determined by considering the above specified macro-system in thermodynamic equilibrium with thermal radiation. In this case the micro-systems populate the different states according to Boltzmann statistics:

$$\frac{N_2}{N_1} = e^{-\frac{W_2 - W_1}{kT}} = e^{-\frac{\hbar\omega_{12}}{kT}} \tag{60}$$

Here $k = 1{\cdot}380 \times 10^{-23}$ W sec/°K is Boltzmann's constant and T the absolute temperature. The distribution according to Boltzmann statistics is shown in Fig. 3. This distribution is obtained from considerations of probability. From all macroscopically identical distributions the Boltzmann distribution may be realized by the largest number of

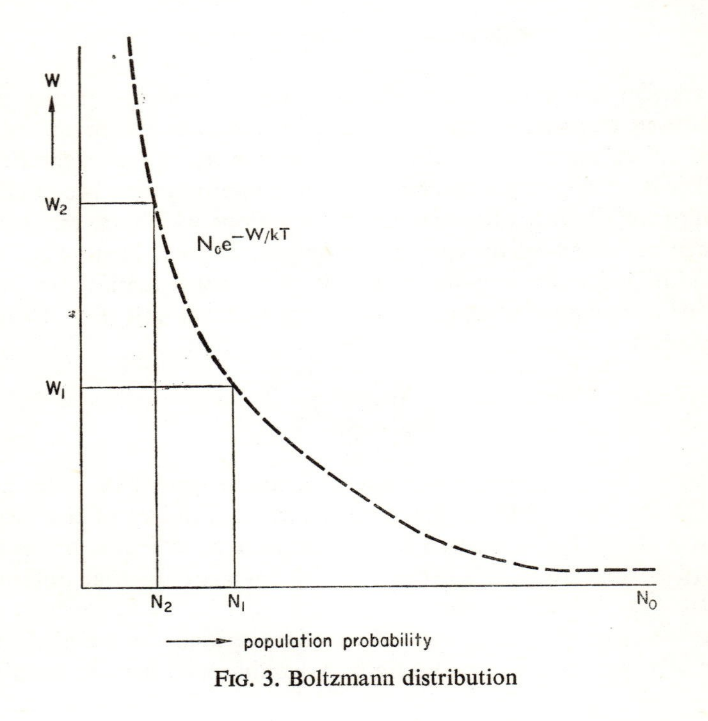

FIG. 3. Boltzmann distribution

microscopically different distributions. It is therefore the most probable distribution in a macro-system which has a large number of micro-systems. When the number of micro-systems is sufficiently large this distribution is so much more probable than all other distributions that it may be considered to be the actual distribution of micro-systems among the various states. According to the Boltzmann distribution the states with higher total energy W are less populated.

The number N of micro-systems in a particular state decreases exponentially with its total energy.

The Boltzmann distribution is now substituted into the condition (59) for equilibrium. N_1 and N_2 are thereby eliminated. Solving for A_{12} we obtain

$$A_{12} = \sigma(\omega_{12})\, B_{12} \left(e^{\frac{\hbar\omega_{12}}{kT}} - 1\right) \tag{61}$$

According to this relation the Einstein coefficient A for spontaneous emission is directly proportional to the corresponding Einstein coefficient B for induced emission. The remaining factor is given by the spectral energy density of thermal radiation. We will therefore next consider thermal radiation in more detail.

1.11. Thermal radiation

Thermal radiation will be studied more thoroughly here because we need its spectral energy density to evaluate eqn. (61) for spontaneous emission. In addition, however, thermal radiation is also of significance in connection with noise in lasers and masers.

To analyse this thermal radiation we refer to a cavity of cubic shape and with sides of length L. The walls of this cavity are assumed to be held at a constant temperature T. As in section 1.3 we will again derive the radiation field from a vector potential $\mathbf{A}$. $\mathbf{A}$ is a solution of the vector wave equation. We can represent the distribution of the field at each instant in time by a Fourier series with period L, in terms of the Cartesian coordinates describing the cubic cavity. We are interested only in that domain L which corresponds to the interior of the cavity. The component A_x of vector $\mathbf{A}$ is thus expressed by the following triple spatial Fourier expansion

$$A_x(\mathbf{r}, t) = \sum_{l,\, m,\, n = -\infty}^{\infty} A^{(x)}_{l,\, m,\, n}(t)\, e^{j\mathbf{k}\cdot\mathbf{r}} \tag{62}$$

The Fourier coefficients are functions of time. Let us assume that these coefficients have a harmonic time dependence according to $\exp(j\omega t)$. These coefficients may then be considered the spectral components of a more general time dependence. For such a spectral component of frequency ω the vectors $\mathbf{k}$ in eqn. (62) have the meaning of propagation vectors with Cartesian components

$$k_x = \frac{2\pi l}{L}, \qquad k_y = \frac{2\pi m}{L}, \qquad k_z = \frac{2\pi n}{L} \tag{63}$$

To be a solution of the wave-equation each individual Fourier component of the field distribution in space with its harmonic time dependence $\exp(j\omega t)$ must satisfy the separation condition

$$k_x^2 + k_y^2 + k_z^2 = k^2 = \frac{\omega^2}{c^2} \tag{64}$$

If the summation in the Fourier expansion is performed independently over all l, m and n each combination of k_x, k_y, and k_z, that is each spatial Fourier component, has a fixed frequency which is specified by eqn. (64). In order to satisfy Maxwell's equations the respective Fourier component must oscillate with this frequency. These Fourier components will therefore be called modes of oscillation. They are the modes of oscillation of the volume $V = L^3$.

With each of these modes $\mathbf{k}$ there may, according to eqn. (63), be associated an element of volume

$$\Delta V_k = \Delta k_x \Delta k_y \Delta k_z = \left(\frac{2\pi}{L}\right)^3 \tag{65}$$

in the space of vector $\mathbf{k}$. The number N_k of modes which are associated with a sphere of radius k in k-space is obtained when the spherical volume $V_k = \frac{4\pi}{3} k^3$ is divided by the volume element ΔV_k from (65):

$$N_k = 2\frac{V_k}{\Delta V_k} = \frac{k^3 L^3}{3\pi^2} = \frac{\omega^3 L^3}{3\pi^2 c^3} \tag{66}$$

The factor 2 in this expression accounts for all modes of oscillation which have orthogonal polarization with respect to the modes from (62). These other modes are either polarized in y-direction or in z-direction.

The total number of modes with wave numbers in the range $0 \ldots k$ or with frequencies between 0 and ω is proportional to the volume of the cavity. If this total number of modes is divided by the volume we obtain

$$n_k = \frac{N_k}{V} = \frac{\omega^3}{3\pi^2 c^3}$$

n_k is the number of modes per unit volume, or the mode density, in the spectral range between 0 and ω. The number of modes per unit volume and per cycle or the spectral mode density is given by

$$n_f = 2\pi \frac{dn_k}{d\omega} = \frac{2\omega^2}{\pi c^3} \tag{67}$$

We have added a factor 2π here because we do not want the spectral density with respect to ω, but with respect to $f = \dfrac{\omega}{2\pi}$.

To find the spectral energy density n_f must be multiplied by the average energy per mode. According to the law of equipartition in statistical mechanics we have $W_m = kT$ for thermodynamic equilibrium of the radiation field with the walls of the cavity at temperature T.

However, this equipartition law is only a limiting case for the statistical distribution at very low frequencies or very high temperatures. This limiting case does not account for the quantum nature of the electromagnetic field. The expression

$$\sigma_0(f) = n_f kT \tag{68}$$

for spectral energy density is therefore correct only for low frequencies as well as for high temperature. The more general law for the distribution will now be obtained by comparing eqn. (61) with (68). The factor $\sigma(\omega_{12})\left[e^{\frac{\hbar\omega_{12}}{kT}} - 1\right]$ in eqn. (61) must be independent of temperature, for A_{12} expresses the probability of a spontaneous process. The probability of any spontaneous process is independent of environmental influences such as the temperature T. Therefore we must have

$$\sigma(f)\left[e^{\frac{hf}{kT}} - 1\right] = \text{const.}$$

and this for any transition frequency $f_{12} = f$. The constant quantity may here be obtained from the limiting case $T \to \infty$. For this limit we have

$$\sigma(f) = \sigma_0(f) = n_f kT$$

while

$$\lim_{T \to \infty} \left(e^{\frac{hf}{kT}} - 1\right) = \frac{hf}{kT}$$

Hence, in general, the spectral energy density is

$$\sigma(f) = n_f \frac{hf}{e^{\frac{hf}{kT}} - 1}$$

and the average energy per mode is

$$W_m = \frac{hf}{e^{\frac{hf}{kT}} - 1} \tag{69}$$

This is the general law for the distribution of energy which takes into account the quantum nature of the radiation field and of matter. Furthermore

$$\sigma(f) = 8\pi \frac{hf^3}{c^3} \frac{1}{e^{\frac{hf}{kT}} - 1} \tag{70}$$

is the commonly used form for the spectral energy density. It is called Planck's radiation law.

With the spectral energy density according to Planck's law the Einstein coefficient for spontaneous emission may now be evaluated. The result is

$$A_{12} = \frac{8\pi h f_{12}^3}{c^3} B_{12} \tag{71}$$

The mean lifetime in state 2, as limited by spontaneous transitions from 2 to 1, is the inverse

$$\tau_{12} = \frac{1}{A_{12}} = \frac{c^3}{8\pi h f_{12}^3 B_{12}} \tag{72}$$

This lifetime decreases proportionally to the inverse of the third power of the transition frequency f_{12}. During a spontaneous transition a photon of frequency f_{12} and energy hf_{12} is emitted. This photon is added with random phase and amplitude to any radiation field which may also be present.

1.12. Non-radiative transitions

Besides induced and spontaneous transitions there is a third type of transition between states which is significant in quantum electronics. These are transitions not connected with any electromagnetic radiation, neither with emission nor with absorption. The energy difference between states in these transitions appears or disappears instead as thermal agitation or heat motion of the surrounding matter. In gases such transitions occur mainly in inelastic collisions. In the solid state, for example a crystalline structure, such transitions are produced by interaction with the vibrating crystal lattice. They are different from induced and spontaneous transitions in that no electromagnetic energy is released or absorbed in these transitions. Rather, the total energy of thermal agitation in gases and liquids, or the number of phonons in the lattice vibrations of crystals changes by the corresponding energy difference.

To account for these non-radiative transitions quantitatively their occurrence is again expressed by transition probabilities. Let s_{12} be the probability for an atom or molecule to make a non-radiative transition from state 2 to state 1 per unit time. The opposite process will be much less probable because, according to Boltzmann statistics, fewer particles or photons with the required thermal energy are available for such energy exchanges. From the distribution given by eqn. (60) it follows that the opposite transition rate is

$$s_{12} = s_{21}\, e^{\frac{\hbar\omega_{12}}{kT}} \tag{73}$$

Actually, to calculate the probabilities for non-radiative transitions would be much more difficult than for induced or spontaneous transitions. Quite a number of factors must be taken into account here. First of all these non-radiative transitions are determined by the composition and characteristics of the surrounding material. For gas lasers or masers the surrounding matter consists of all the components of the gas mixture, in the case of solid state lasers or masers it is the lattice of the host crystal.

These few remarks about non-radiative transitions will suffice for the present. More details will be given later when specific examples of such transitions are considered.

1.13. Linewidth of induced transitions

The probability for induced transitions of a micro-system resulting from interaction with an electromagnetic field of harmonic time dependence is given by eqn. (51) and displayed in Fig. 1 as a function of frequency. If the interaction lasts sufficiently long the frequency characteristic degenerates into a very sharp resonance peak. From this it appears that only when the frequency of the oscillating field is exactly tuned to the transition frequency

$$f_{12} = \frac{W_2 - W_1}{h}$$

will there be induced transitions between states 1 and 2. For the slightest difference between these frequencies transitions seem to be less and less probable compared with the rate at resonance, the longer the interaction lasts.

From the preceding three sections we have learned that interaction between an electromagnetic field and a micro-system in a definite state can never last for an arbitrarily long time. States of the micro-system from which spontaneous transitions can be made have a limited lifetime. The average lifetime due to spontaneous transitions is determined by eqn. (72). In addition non-radiative transitions will further shorten the lifetime in definite states. The time of interaction between an electromagnetic field and the micro-system will thus be more or less limited. This limitation will change the frequency characteristic of transition probability from that which is given by eqn. (51).

To study the effects of the limited lifetime we again refer to a macro-system consisting of a large number of micro-systems which are all independent of one another. The micro-systems are assumed to have states 1 and 2 with energy levels as shown in Fig. 2. The number of spontaneous transitions from 2 to 1 is directly proportional to the number N_2 of micro-systems in state 2. Therefore the population of this state decays exponentially with time through spontaneous transitions, very much as in the case of radioactive decay. The time constant of this exponential function is the mean lifetime

$$\tau_{12} = \frac{1}{A_{12}}$$

If at time t_0 a number $N_2(t_0)$ of particular micro-systems is in state 2 then at a time $t > t_0$ only a number

$$N_2(t-t_0) = N_2(t_0)\, e^{-(t-t_0)/\tau_{12}}$$

of these same micro-systems will have remained in state 2. There may be other micro-systems of the macroscopic arrangement which, in the time interval between t_0 and t, have made a transition into state 2. But over the whole time interval from t_0 to t only these $N_2(t-t_0)$ micro-systems have always remained in state 2.

If, on the other hand, at time t a number $N_2(t)$ of definite micro-systems are in state 2 only

$$N_2(t-t_0) = N_2(t)\, e^{-(t-t_0)/\tau_{12}}$$

of them were in the same state 2 at the previous time t_0. Only this fraction $N_2(t-t_0)$ of the $N_2(t)$ micro-systems has always been in state 2 since time t_0.

According to eqn. (51) the transition probability depends on the time interval during which a particular micro-system has been in one and the same state, because only during this time can the interaction with the field be described by the respective matrix element. During the time interval dt_0 at t_0 the increment

$$dN_2(t-t_0) = \frac{\partial N_2(t-t_0)}{\partial t_0}\, dt_0 = \frac{N_2}{\tau_{12}}\, e^{-(t-t_0)/\tau_{12}}\, dt_0$$

is added to the number $N_2(t)$ of micro-systems which, at time t, are still in state 2.

At time t

$$w_{12}(t)\, dN_2(t-t_0) = \frac{2}{\hbar^2}\, |\mathbf{E}_0 \cdot \mathbf{P}_{12}|^2\, \frac{\sin^2 \frac{1}{2}(\omega-\omega_{12})(t-t_0)}{(\omega-\omega_{12})^2} \times$$

$$\times \frac{N_2}{\tau_{12}}\, e^{-(t-t_0)/\tau_{12}}\, dt_0$$

of the increment $dN_2(t-t_0)$ of micro-systems has been stimulated to make a transition to state 1. Let us now integrate this expression over all time from $t_0 = -\infty$ to t. The result is:

$$N_{12} = \frac{2}{\hbar^2} |\mathbf{E}_0 \cdot \mathbf{P}_{12}|^2 \frac{N_2}{\tau_{12}(\omega-\omega_{12})^2} \times$$

$$\times \int_{-\infty}^{t} e^{-(t-t_0)/\tau_{12}} \sin^2 (\omega-\omega_{12})(t-t_0)\, dt_0$$

N_{12} is the number of micro-systems which, while continuously interacting with the harmonically oscillating field, have made transitions from state 2 to state 1. When making the substitution $u = t-t_0$ for the variable of integration in the above expression a definite integral results which may be evaluated, i.e.

$$\int_0^{\infty} e^{-\frac{u}{\tau_{12}}} \sin^2 \tfrac{1}{2}(\omega-\omega_{12})\, u\, du = \frac{1}{2} \frac{\tau_{12}^3(\omega-\omega_{12})^2}{1+\tau_{12}^2(\omega-\omega_{12})^2}$$

We thus obtain

$$N_{12} = \frac{1}{\hbar^2} |\mathbf{E}_0 \cdot \mathbf{P}_{12}|^2 \frac{N_2}{(\omega-\omega_{12})^2 + \frac{1}{\tau_{12}^2}} \tag{74}$$

The number N_{12} is independent of time t. While continuously interacting with the electromagnetic field a stationary state is established for the macro-system. Under this equilibrium we have N_{12} less micro-systems in state 2 than would be the case when no field interaction is taking place. At the same time the number of micro-systems in state N_1 has been increased by this number N_{12}. The same electromagnetic field also stimulates transitions from 1 to 2. The transition probabilities are equal for both directions. We therefore have

$$N_{21} = N_{12} \frac{N_1}{N_2} \tag{75}$$

where the meaning of N_{21} follows from that of N_{12}.

Let us designate the equilibrium populations of states 1 and 2 under no field interaction by $N_1^{(e)}$ and $N_2^{(e)}$ respectively. Then population of each state when the electromagnetic field does interact with the micro-system is

$$N_2 = N_2^{(e)} - N_{12} + N_{21}$$
$$N_1 = N_1^{(e)} - N_{21} + N_{12}$$

Using eqns. (74) and (75) we obtain for the difference in population

$$N_1 - N_2 = \frac{N_1^{(e)} - N_2^{(e)}}{1 + 2\Lambda} \tag{76}$$

Here we have written

$$\Lambda = \frac{1}{\hbar^2} |\mathbf{E}_0 \cdot \mathbf{P}_{12}|^2 \frac{1}{(\omega - \omega_{12})^2 + \dfrac{1}{\tau_{12}^2}} \tag{77}$$

for convenience.

When interaction with an electromagnetic field occurs, the difference in population will always be smaller than for equilibrium with no field present. If the interaction is sufficiently weak then $\Lambda \ll 1$ and

$$N_1 - N_2 \simeq (N_1^{(e)} - N_2^{(e)})(1 - 2\Lambda)$$

Now we have

$$\Delta N = (N_1^{(e)} - N_2^{(e)})\Lambda$$

more micro-systems in state 2 than there would be at equilibrium. Due to spontaneous transition from 2 to 1 they have a mean lifetime τ_{12}. Hence $\Delta N/\tau_{12}$ of these ΔN micro-systems will spontaneously change back to state 1 per unit time. For a stationary state of the macro-system the difference ΔN will, however, be maintained constant. Therefore the interacting electromagnetic field must continuously supply the following power

$$P = \frac{\Delta N}{\tau_{12}} \hbar\omega_{12} = \frac{N_1^{(e)} - N_2^{(e)}}{\tau_{12}} \Lambda\hbar\omega_{12} \tag{78}$$

This relation for power still holds when an equilibrium distribution of states $N_1^{(e)}$ and $N_2^{(e)}$ is created which differs from the Boltzmann distribution due to other external influences. The power is still given by eqn. (78) even when $N_1^{(e)} - N_2^{(e)} < 0$. It will be negative under these conditions, meaning that the induced emission exceeds the absorption and that, on average, power will be released from the micro-systems and added to the interacting electromagnetic field.

The frequency characteristic is the same whether the net effect is induced absorption or induced emission. From eqn. (77) this characteristic is given by the factor

$$g_0(\omega) = \frac{2/\tau_{12}}{(\omega - \omega_{12})^2 + 1/\tau_{12}^2} \tag{79}$$

By including $2/\tau_{12}$ this factor has been normalized so that

$$\int_0^\infty g_0(f)\,df = \frac{1}{2\pi}\int_{-\infty}^{\infty} g_0(\omega)\,d\omega = 1 \tag{80}$$

Under this condition $g_0(f)$ is called the normalized line shape. More generally any frequency characteristic of the induced emission or absorption is called the normalized line shape if it is normalized according to eqn. (80).

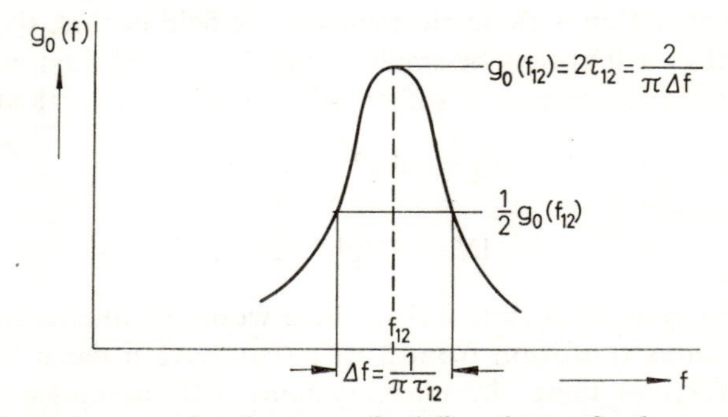

FIG. 4. Lorentz curve for the normalized line shape of a homogeneously broadened line due to spontaneous transitions

The specific line shape of eqn. (79), as it is effected by the limited lifetime of micro-systems due to spontaneous transitions, is called the Lorentz curve. This characteristic is shown in Fig. 4. The half-value or 3-db bandwidth is

$$\Delta f = \frac{1}{\pi \tau_{12}} \tag{81}$$

It is called the natural linewidth.

The emission or absorption line for an unlimited lifetime and long interaction is extremely narrow. Due to the limited lifetime it is effectively broadened so that purely sinusoidal field oscillations induce transitions even if their frequency is detuned somewhat from f_{12}. When, as in our present analysis, the lifetime in participating states of the micro-systems is short compared with the interaction time, the effect is called *homogeneous line broadening*. In this case the participation of micro-systems in induced transitions is homogeneously distributed over all micro-systems.

Using the normalized line shape the power which is absorbed or emitted during stimulated transitions may be written as follows:

$$P = \frac{1}{2\hbar}\omega_{12}|\mathbf{E}_0 \cdot \mathbf{P}_{12}|^2(N_1 - N_2)\, g_0(f) \tag{82}$$

Here the effects of spontaneous transitions and of the mean lifetime τ_{12} are all included in $g_0(f)$. No other factor of eqn. (82) depends directly or indirectly on τ_{12}. This already shows that eqn. (82) is of more general significance and not just valid for the specific line shape of eqn. (79).

Actually there are a number of other phenomena which also influence the broadening of the emission line, often much more than the broadening effect of spontaneous emission. The line is broadened, for example, when the energy levels in Fig. 2 of participating micro-systems are scattered randomly about an average level. When a host crystal has been doped with active micro-systems such deviation of energy levels from an average value is caused by interaction of the active micro-systems with the host lattice. If the active micro-systems are in a gaseous state, or one component of a mixture of gases, the random heat motion will cause a scattering of the energy levels by the Doppler effect. When the energy levels of the individual micro-systems are scattered, the corresponding transition frequencies will deviate from their nominal values.

Let this deviation be described by a distribution function $g(f_{12})$. Of all $(N_1 - N_2)$ micro-systems which, according to eqn. (82), are participating in absorption or emission of power the incremental number

$$d(N_1 - N_2) = (N_1 - N_2)\, g(f_{12})\, df_{12}$$

will be in the interval df_{12} at f_{12}. The integration of $d(N_1 - N_2)$ over the full range of frequencies must result in $N_1 - N_2$, hence we have

$$\int_0^\infty g(f_{12})\, df_{12} = 1 \tag{83}$$

The field which is interacting with all micro-systems will induce in particular, for the fraction $d(N_1 - N_2)$ of the micro-systems, the following emission or absorption of power

$$dP = \frac{1}{2\hbar}\omega_{12}|\mathbf{E}_0 \cdot \mathbf{P}_{12}|^2(N_1 - N_2)\, g(f_{12})\, g_0(f)\, df_{12}$$

Integrating over all micro-systems the total power which is absorbed or emitted at frequency f of the stimulating field is

$$P = \frac{\pi}{\hbar} |\mathbf{E}_0 \cdot \mathbf{P}_{12}|^2 (N_1 - N_2) \int_0^\infty f_{12} g(f_{12}) g_0(f) \, df_{12}$$

To evaluate the integral we note that normally the distribution given by $g(f_{12})$ is much broader than the relatively sharp line $g_0(f)$ due to spontaneous transitions. $g_0(f)$ is quite often sufficiently sharp that when integrating we may represent $g_0(f)$ by Dirac's function:

$$g_0(f) = \delta(f_{12} - f)$$

Under such conditions we obtain

$$P = \frac{\pi}{\hbar} |\mathbf{E}_0 \cdot \mathbf{P}_{12}|^2 (N_1 - N_2) f g(f) \tag{84}$$

Comparing this expression with eqn. (82) the distribution function $g(f)$ is identified as the effective line shape of the absorption or emission line. The shape of this line is now solely determined by the scattering of transition frequencies. Because (83) is satisfied, $g(f)$ is again a normalized line shape, as was $g_0(f)$.

When a monochromatic oscillation is interacting with micro-systems whose transition frequencies deviate randomly from an average value, the interaction is not equally effective for all micro-systems. The micro-systems which are close in their transition frequency to the frequency of external field will be stimulated more often to make transitions, than the other systems. This is in contrast to stimulated transitions for short lifetime, and long interaction where the effect of interaction is equally strong on all micro-systems.

While the line broadening was called homogeneous for equally effective interaction with all micro-systems, it is called *inhomogeneous line broadening* when the interaction is selective for a particular fraction of the micro-systems.

The power which is absorbed or emitted during transitions which are stimulated by a sinusoidally oscillating field is always given by the same expression whether it be homogeneous or inhomogeneous line broadening. We must only substitute for $g(f)$ the normalized line shape which describes the particular situation correctly.

For any such stimulated transition of a particular micro-system $g(f)$ is a statistical quantity. $g(f)df$ is the probability that the transition between states 1 and 2 will interact with a photon in the frequency interval between f and $f+df$.

The normalization

$$\int_0^\infty g(f)\, df = 1$$

accounts for the fact that for each stimulated transition a photon is either absorbed or emitted.

CHAPTER 2

The Laser

THE laser utilizes stimulated emission either to amplify or to generate electromagnetic waves and signals in the optical range of frequencies. Optical frequencies are in a spectral range for which the dimensions of normal physical arrangements are large compared with the free space wavelength of electromagnetic oscillations. This is in contrast to the maser. The maser is utilized for amplification of microwaves. Here the dimensions of the physical arrangements are of the same order of magnitude as the free space wavelength. Because the geometric dimensions of lasers are large compared with the wavelength their discussion is in many respects not as difficult as the discussion of masers. We will therefore continue here by treating the laser first. This order is also justified in that there are likely to be many more technical applications of the laser than for the maser.

To obtain induced emission the population of certain energy levels must be inversely distributed compared with the normal distribution at thermal equilibrium. The difference in population $(N_1 - N_2)$ in eqn. (84) must be negative. We will therefore discuss first how to obtain this inversion in population, and which combination of energy levels will be needed for this inversion.

2.1. Population inversion

In thermal equilibrium the different states are populated according to Boltzmann statistics (60). In this case most of the atoms or molecules are in states of lower energy and the higher the energy of a particular state the fewer micro-systems will populate it. The distribution is shown in Fig. 3.

To obtain an inverted population energy of a certain kind must be supplied in a specific manner. This energy is called pumping energy. Depending on the particular laser medium energy may, for example, be supplied by absorbing electromagnetic radiation. This radiation must be of a higher frequency than the oscillation which is to be amplified by stimulated emission. The pumping radiation may, however, be incoherent and even be distributed over a relatively wide spectral range. Figure 5 shows which energy levels of the micro-systems participate in this process.

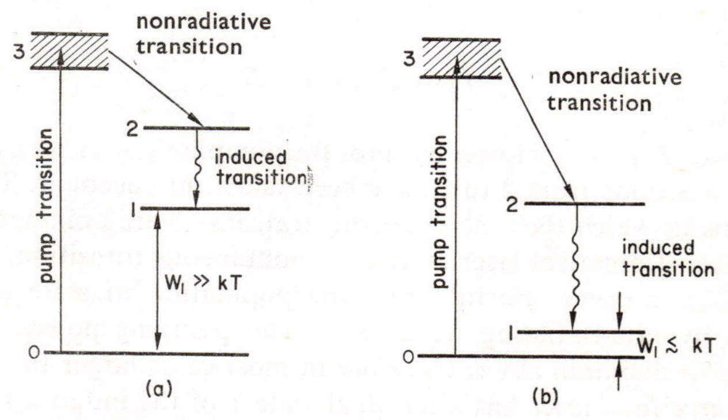

FIG. 5. Energy levels in lasers: (a) four-level laser; (b) three-level laser

By irradiating the medium or otherwise supplying pumping energy a large number of micro-systems is excited from the ground state 0 into states within the energy band 3. From here a sufficient number of these excited micro-systems must make non-radiating transitions to state 2. State 2 should be metastable; its lifetime should be sufficiently long. It should at any rate be larger than the lifetime of the micro-systems in state 1. Thus the population N_2 of state 2 can become larger than the population N_1 of state 1.

When specifying the pumping energy which must be supplied to achieve population inversion for stimulated emission two different classes of laser materials must be distinguished according to the arrangement of participating energy levels. The corresponding energy level diagrams are shown in Fig. 5.

The *four-level laser* has a terminal induced transition state 1 for which the energy is $W_1 \gg kT$. This energy level is sufficiently high above the ground state for its equilibrium population N_1, according to Boltzmann statistics, to be much smaller than the population inversion

$\Delta N = N_1 - N_2$ which is needed to obtain significant amplification by stimulated emission. With $N_1 \ll \Delta N$, if ΔN is specified for such amplification, we only need to supply sufficient pumping energy for the population of state 2 to grow to

$$N_2 \simeq \Delta N \tag{85}$$

$A_{12}N_2$ of the micro-systems make spontaneous transitions from 2 to 1 per unit time. Hence the power to be supplied to maintain the population inversion ΔN must be at least

$$P_f = A_{12}\hbar\omega_{03}\Delta N \tag{86}$$

This power P_f is only a lower limit for the pumping power. Only spontaneous transitions from 2 to 1 have been taken into account. There are situations in which these spontaneous transitions are dominant. However, in most practical laser materials spontaneous transitions to other states than 1 occur. Furthermore the population of state 2 will be drained by non-radiating transitions. The pumping power which is required to maintain ΔN is therefore in most cases larger than (86).

The *three-level laser* has a terminal state 1 of the induced transition which is energetically only just above the ground state. With its energy $W_1 < kT$ this state is, according to Boltzmann statistics, heavily populated even at equilibrium. Most of the pumping energy must first be supplied only to populate state 2 through the band of states 3 so that N_2 will be as large as N_1. The population inversion which is required in addition to achieve amplification by induced emission is then only small compared with the total population N_2 of 2. We therefore have $N_1 \gg \Delta N$, and in order to get laser action it is required that

$$N_2 \simeq N_1 \tag{87}$$

The three-level laser requires much more pumping power than the four-level laser. The ratio of the pumping powers in each class is nearly $N_2/\Delta N$. In the case of the three-level laser a significant part of all N active micro-systems is in state 1 at equilibrium. In order to obtain induced emission nearly half of all these active micro-systems must be excited into state 2. For typical laser materials the ratio of N_2 for population inversion is $N_2/\Delta N \simeq 100$. Three-level lasers must usually be pumped with 100 times the pumping powering of four-level lasers.

2.2. Amplification

Let a medium contain micro-systems with states 1 and 2, between which an external electromagnetic oscillation is able to induce transitions. If the populations of these two states are inverted induced emission in the medium will be larger than absorption. Under these conditions, according to eqn. (84) energy will be released from the micro-systems to the external oscillation. The oscillation will be amplified.

To calculate the amplification we will rewrite eqn. (84) for the emitted power in a way which is similar to eqn. (54) for the interaction with thermal radiation:

$$P = \frac{hf}{c} B_{12}(N_2 - N_1)\, g(f)\, S \tag{88}$$

Here we have written $S = |E_0|^2/\eta$ for the radiation density of the exciting wave and with

$$B_{12} = \frac{2\pi^2}{h^2 \varepsilon_0} |\mathbf{P}_{12}|_E^2 \tag{89}$$

we have defined a new Einstein coefficient for those transitions which are induced by a linearly polarized monochromatic field. $(\mathbf{P}_{12})_E$ is the matrix element (1, 2) for that component of the electric dipole-moment of the micro-system which is parallel to the electric field vector $\mathbf{E}_0$. We will henceforth express the emitted power as in eqn. (88) following common practice.

Let us apply eqn. (88) to the arrangement in Fig. 6. The homogeneous laser material of cylindrical volume V is assumed to be excited by uniformly pumping it to a constant population inversion throughout the volume. For an element dV of volume V eqn. (88) gives

$$\frac{dP}{dV} = \frac{hf}{c} B_{12}(n_2 - n_1)\, g(f)\, S$$

Here $n = dN/dV$ is the population density of a particular state. $(n_2 - n_1)$ is the inversion density.

Consider an incident electromagnetic wave which is plane and uniform and travelling in the z-direction. By stimulated emission its radia-

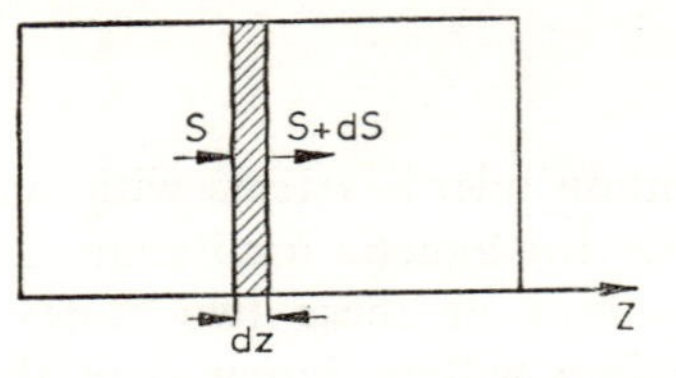

FIG. 6. Homogeneous laser medium with uniform population inversion

tion density changes according to

$$\frac{dS}{dz} = \frac{hf}{c} B_{12}(n_2 - n_1) g(f) S$$

The general solution of this differential equation for S is

$$S(z) = S(0) e^{2vz}$$

where

$$v = \frac{hf}{2c} B_{12}(n_2 - n_1) g(f) \tag{90}$$

This solution shows that the radiation density increases exponentially. v is the logarithmic amplification or gain per unit length for the wave amplitude. This gain is proportional to the Einstein coefficient B_{12}, the inversion density, and the normalized line shape at the frequency f of the incident wave.

To evaluate eqn. (90) further the Einstein coefficient B_{12} needs to be determined. It is out of question to try and calculate this coefficient through the relations for the matrix elements for most practical situations. For active micro-systems in most laser materials such calculations would be very cumbersome if not impossible. Very often not even the specific state functions are known accurately enough.

Instead of a computation the Einstein coefficient may be measured. To this end we let the wave be incident on the laser medium without any pumping power for inversion of the population. Under these conditions $n_2 < n_1$ and their ratio is given by the Boltzmann distribution. The medium now absorbs power instead of emitting it. The wave is attenuated instead of amplified. The attenuation constant is

$$\alpha = \frac{hf}{2c} B_{12}(n_1 - n_2) g(f)$$

If α is measured at all frequencies within the absorption line and then integrated the result is

$$\alpha_i = \int \alpha \, df = \int \frac{hf}{2c} B_{12}(n_1 - n_2) \, g(f) \, df$$

With $\int g(f) \, df = 1$, and neglecting the gradual change due to the frequency factor f compared with $g(f)$ the following formula results for the Einstein coefficient:

$$(n_1 - n_2) B_{12} = \frac{2c}{hf_{12}} \alpha_i \tag{91}$$

In addition this measurement gives all the details of the normalized line shape.

The effect of the normalized line shape on the amplification of the medium will be known when the gain formula (90) is evaluated at the frequency of maximum gain and when in addition the 3-db bandwidth is substituted into (90). For this evaluation specific assumptions about the shape of the emission line $g(f)$ must be made.

Let us first assume purely homogeneous line broadening by spontaneous transitions with long-lasting interaction. In this case the normalized line shape is a Lorentz curve

$$g_0(f) = \frac{2}{\pi \Delta f} \frac{1}{1 + \left(2 \dfrac{f - f_{12}}{\Delta f}\right)^2} \tag{92}$$

As a second example we will assume inhomogeneous line broadening by the Doppler effect of thermal motion in gases or by similar random deviations of the transition frequencies about an average value. We will show later in a more detailed discussion of gas lasers that such inhomogeneously broadened lines are described by a Gaussian curve

$$g(f) = \frac{2}{\Delta f} \sqrt{\frac{\ln 2}{\pi}} \exp \left[-\ln 2 \left(2 \frac{f - f_0}{4f} \right)^2 \right] \tag{93}$$

Here f_0 is the arithmetic mean of the transition frequencies.

The maximum value of the Lorentz curve in the case of homogeneously broadened lines is

$$g_0(f_{12}) = \frac{2}{\pi \Delta f}$$

and the maximum value of the Gaussian curve in the case of inhomogeneously broadened lines is

$$g(f_0) = \frac{2}{\Delta f}\sqrt{\frac{\ln 2}{\pi}}$$

Consequently the maximum gain for these two cases is different. For a Lorentz curve we have

$$v(f_{12}) = \frac{hf_{12}}{\pi c\Delta f} B_{12}(n_2 - n_1) \tag{94}$$

while for a Gaussian curve we have

$$v(f_0) = \sqrt{\frac{\ln 2}{\pi}} \frac{hf_0}{c\Delta f} B_{12}(n_2 - n_1) \tag{95}$$

For both line shapes the maximum gain is proportional to the inverse of the 3-db bandwidth. This relation holds quite generally for any line shape. Only the constant of this inverse proportionality depends on the specific line shape.

If the laser is directly utilized as an amplifier its power gain and frequency characteristic are significant. We need power gain over a sufficiently wide band of frequencies to amplify signals which contain

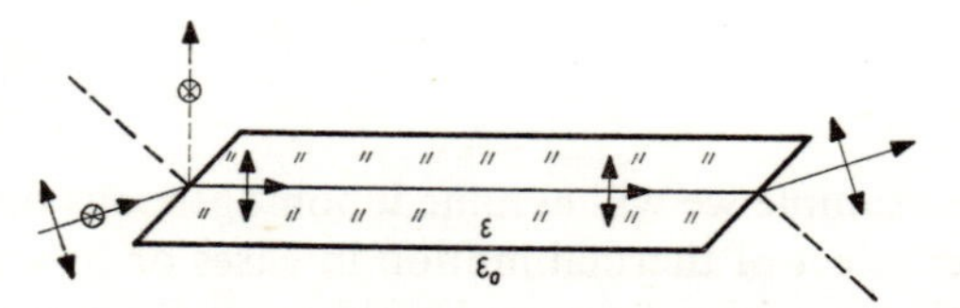

FIG. 7. Light amplifier with input and output surfaces under the Brewster angle

a high amount of information. Power gain of an amplifier is defined as the ratio of output power to the power which is available at the input of the amplifier.

Let us consider the basic arrangement of Fig. 7 for a laser amplifier.

A uniform plane wave, linearly polarized in the plane of Fig. 7, is assumed to be incident on the system. Its direction of propagation is so adjusted that it makes the Brewster angle with the oblique input and output surfaces. Under these circumstances there will be no reflection at either surface, and the actual input power will be equal to the available power at the input. If the laser has length l in the direction of transmission its power gain is

$$V = e^{2v(f)l} \tag{96}$$

Designating the maximum gain at the centre frequency f_0 of the band by V_0 we have

$$\ln V_0 = 2v(f_0)\, l$$

The limits of the band $f = f_0 \pm \delta f/2$ will be defined as the frequencies where the power gain has dropped to -3 db of its maximum value V_0. They are obtained from

$$2v\left(f_0 \pm \frac{\delta f}{2}\right) l = \ln V_0 - \ln 2$$

Solving this equation for δf in case of the Lorentz curve results in

$$\delta f = \Delta f \sqrt{\frac{\ln V_0}{\ln V_0 - \ln 2} - 1} \tag{97}$$

while for a Gaussian curve the result is

$$\delta f = \Delta f \sqrt{\frac{\ln \dfrac{\ln V_0}{\ln V_0 - \ln 2}}{\ln 2}} \tag{98}$$

Both results for the half-power bandwidth have been divided by the 3-db bandwidth Δf of the emission line, and plotted in Fig. 8 against the maximum power gain at the centre of the band.

From this diagram we see that the half-power bandwidth decreases with increasing power gain. Such narrowing of the bandwidth is typical for all travelling wave amplifiers. It is caused by the frequency dependence of the gain constant. Signal amplitudes grow faster exponentially at the centre than they do at the limits of the band. When the power gain at the centre of the band is 20 db the half-power bandwidth has been narrowed to less than half the 3 db bandwidth of the emission line. This decrease in the bandwidth which depends on the power gain is nearly equal for both line shapes. For small power gain the half-power bandwidth becomes larger than the 3-db bandwidth of the emission line. At $V_0 = 2$ db it tends to infinity. For this value of the power gain at the centre of the band no amplification is needed for the half-power points provided the laser is transparent at all frequencies.

Another significant characteristic of amplifiers is the saturation of amplification. So far we have only calculated power gain under the assumption that interaction between waves and matter is weak everywhere within the laser medium. In eqn. (78) we have let $\Lambda \ll 1$, thus assuming that the number of microsystems ΔN in (78) which have been induced by the incident signal into transitions from their zero signal

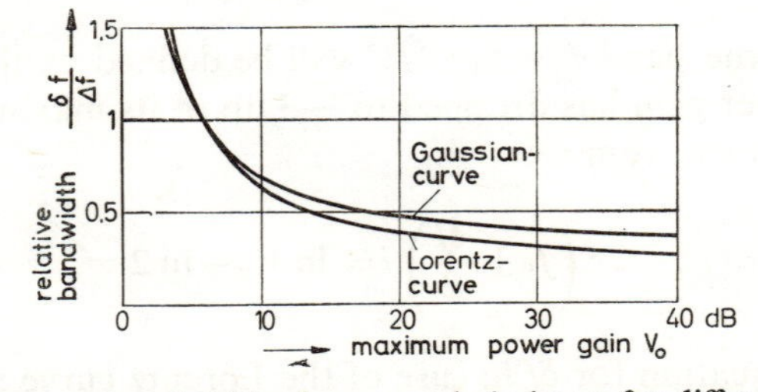

FIG. 8. Half-power bandwidth of power gain in lasers for different line shapes

equilibrium populations $N_1^{(e)}$ and $N_2^{(e)}$ is small compared with $N_1^{(e)} - N_2^{(e)}$. The power gain which has been obtained under these conditions is independent of signal amplitude. The amplification has the typical small signal characteristics where the output amplitude increases linearly with the input amplitude.

Now for large signals, starting at a certain signal amplitude, the population inversion will be reduced by stimulated emission sufficiently to change the gain factor of the medium. For very large signals the amplification will eventually reach saturation. Increasing the input amplitude will then no longer increase the output amplitude proportionally.

Quantitative information about saturation may be obtained from eqn. (76). According to this equation the population N_2 of the upper level is changed by continuously interacting with a sine-wave from the equilibrium population corresponding to no incident wave. The change is

$$\Delta N = -\frac{1}{2}\left[\frac{N_1^{(e)} - N_2^{(e)}}{1+2\Lambda} - (N_1^{(e)} - N_2^{(e)})\right] = (N_1^{(e)} - N_2^{(e)})\,\frac{\Lambda}{1+2\Lambda} \qquad (99)$$

This change will become significant compared with $(N_1^{(e)} - N_2^{(e)})$ when $\Lambda \simeq 1$. If values for Λ are of this magnitude the denominator will be so very different from unity that ΔN, as well as the power of the induced emission, are no longer proportional to the intensity of the sine-wave. Furthermore the shape of the emission line will be changed from the Lorentz curve with its natural linewidth. Because of the additional limitation on the lifetimes in state 2 by induced transitions the line will be broadened. The effect is called *saturation broadening* of the natural line. Maximum gain will however still be obtained when resonance with the transition frequency $f = f_{12}$ is achieved.

Noting the dependence of (99) on Λ a saturation limit for the intensity of the exciting field may be defined from $\Lambda = 1$. When eqns. (77), (79) and (88) are taken into account this condition is found to be identical with

$$B_{12} g_0(f) \frac{S}{c} = A_{12} \tag{100}$$

On the right-hand side we have the Einstein coefficient for spontaneous emission, i.e. the spontaneous transition rate per micro-system in state 2. On the left-hand side we have the induced transition rate, again per micro-system in state 2. Hence, according to eqn. (100), the laser medium goes into saturation when the incident wave would cause as many induced transitions from the equilibrium position as there are spontaneous transitions.

Saturation starts at the output end of the amplifier. With the onset of saturation the gain constant will change and become dependent on longitudinal distance. The overall power gain decreases.

The situation is still more involved when the input signal is not constant in time but modulated in amplitude, for example. The rate of change of the amplification of a laser at saturation is limited. To invert the population of two particular states for stimulated emission micro-systems will enter the upper state 2 by non-radiative transitions from other states. These non-radiative transitions are relaxation processes which do not change the population of state 2 at any rate faster than is given by the corresponding relaxation time. Amplification at saturation will therefore change only as fast as the rate determined by this relaxation time. For amplitude-modulated signals with modulation frequencies much higher than the inverse of the corresponding relaxation time the laser will have a constant gain even if it is driven into saturation. No combination frequencies or harmonics of modulation frequencies will be generated in this case. The amplification is still linear with respect to the amplitude modulation. This is called *super-linear amplification.*

Information about the relaxation process may be obtained experimentally by observing the amplification of signal pulses which have very steep sides but a flat top, and which drive the laser into saturation. The rise time should be much shorter than the relaxation time and the flat top longer than τ_R. As is shown in Fig. 9 the wave front will be amplified initially almost linearly according to the small signal gain. Soon after, however, the population of the upper state for induced emission will be depleted. Starting at the output end and receding to

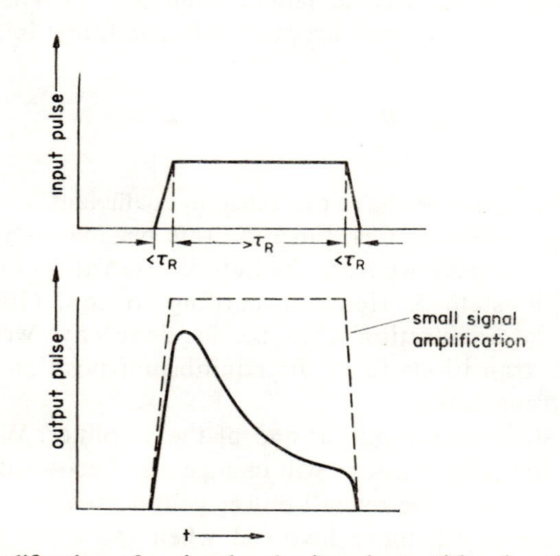

FIG. 9. Amplification of a signal pulse in a laser with relaxation time τ_R

the input the amplification will progressively become saturated. The result is an output pulse which, instead of having a flat roof, decays according to the relaxation process.

2.3. Laser noise

The ultimate sensitivity of an amplifier is limited by noise. The noise characteristics of a laser amplifier are therefore as important as its amplification. For lasers, as well as for masers, we have thermal noise and the noise of spontaneous emission as noise sources. Here we will study the noise effects of spontaneous emission. We consider the laser arrangement shown in Fig. 10.

The power of spontaneous emission per unit volume of laser medium is

$$R_{sp} = A_{12}n_2hf_{12}$$

This power density has a spectral distribution of random frequency components corresponding to the normalized line shape $g(f)$. It is randomly radiated into all directions of space. Therefore the solid angle of radiation is 4π. If the emission line is only broadened by spontaneous transitions $g(f)$ will have the natural line shape $g_0(f)$. If, in addition, the transition frequencies of individual micro-systems deviate from one another the emission line will be further broadened and shaped differently. Let the signal which is to be amplified have a spectrum of width δf. We will assume δf to be narrow compared with the width of $g(f)$. Under these conditions only the noise power R_s within the band of signal frequencies needs to be taken into account.

$$R_s = A_{12}n_2hf_{12}g(f)\,\delta f$$

Furthermore, of the noise power in this limited spectral range we need to consider only the part $\frac{\Delta\Omega}{4\pi} R_s$ which is radiated into the solid angle $\Delta\Omega$ over which the signal wave extends. We obtain

$$R_0 = \frac{1}{2} A_{12}n_2hf_{12}\,g(f)\,\delta f\frac{\Delta\Omega}{4\pi} \tag{101}$$

where the factor $\frac{1}{2}$ takes into account only the polarization parallel to the signal polarization from both polarizations of noise radiation.

The low noise amplifier in Fig. 10 has all the elements which will eliminate all noise components outside the signal band δf and outside the solid angle $\Delta\Omega$ as well as the noise component of the orthogonal polarization with respect to the signal wave. A convex lens following the laser output focuses onto the small iris only those rays within the small solid angle $\Delta\Omega$. These rays are transmitted and subsequently transformed back into a nearly parallel beam by another convex lens. This beam passes through an interferometer formed by two partly transparent parallel plane mirrors. When the distance between the mirrors is half a wavelength or an integral multiple of $\lambda/2$ the arrangement is resonant and becomes transparent. Thus the frequency band of the signal spectrum is filtered out. Finally a polarization analyser eliminates the polarization which is orthogonal to the signal wave.

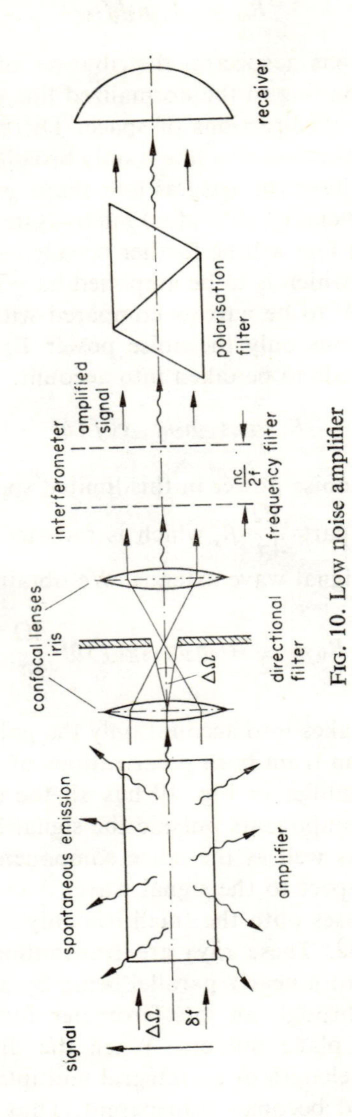

FIG. 10. Low noise amplifier

In eqn. (101) the factor $\delta f \Delta\Omega/4\pi$ may be related to the number of modes of oscillation in the volume of the laser. We have

$$\frac{\delta f}{\Delta f}\frac{\Delta\Omega}{4\pi} = \frac{n_s}{n_f Q_L L \Delta f} \tag{102}$$

where as before Δf is the width of the emission line, while Q_L is the cross-sectional area of the laser and L its length. n_f is the spectral mode density and n_s is the number of modes needed to represent the signal wave of bandwidth δf within the solid angle $\Delta\Omega$. The product $n_f Q_L L \Delta f$ is the total number of modes needed to represent the radiation field of the spontaneous emission in the laser volume. We could also say that only these modes are excited by spontaneous emission. In eqn. (102) we have stated that bandwidth times the solid angle of the signal wave to bandwidth times the solid angle of the spontaneous emission are in the same ratio as the corresponding mode numbers. This is certainly a reasonable statement.

If at all possible we will let the signal wave excite only that mode of propagation in the laser medium which is of lowest order dependence of its fields on the transverse coordinates. In waveguide theory this mode of lowest transverse order is called the dominant mode of propagation. It corresponds most nearly to the uniform plane wave when the medium is not limited in the transverse direction. With only one transverse mode the number of oscillating signal modes n_s will be determined solely by the width δf of the signal band. For a laser of a length L the frequency spacing between adjacent modes of different longitudinal order is $\frac{c}{2L}$. Within the band δf we have therefore

$$n_s = \frac{2L}{c}\delta f \tag{103}$$

modes of the same transverse order. Substituting n_s from (103) into (102) we obtain

$$\delta f \frac{\Delta\Omega}{4\pi} = \frac{2\delta f}{c n_f Q_L}$$

Using this expression in eqn. (101) the noise power density which is added per unit volume to the signal wave within the signal band is

$$R_0 = A_{12}\frac{n_2 h f_{12} g(f)\,\delta f}{c n_f Q_L} \tag{104}$$

When this spontaneous noise power propagates through the laser it will be amplified by stimulated emission. The rate of change of the radiation intensity R of this noise is given by

$$\frac{dR}{dz} = 2vR + R_0 \tag{105}$$

The first term on the right-hand side accounts for the amplification by stimulated emission. The second term represents spontaneous emission. The solution of this differential equation with $R(0) = 0$ as initial condition is

$$R = \frac{R_0}{2v}(e^{2vL} - 1) \tag{106}$$

We substitute from (104) for R_0 in this expression while we substitute for v from (90). If, furthermore, we use (71) to relate A_{12} to B_{12} and (67) for n_f the result is

$$R = \frac{hf_{12}}{Q_L} \frac{n_2}{n_2 - n_1} \delta f(e^{2vL} - 1)$$

This is the radiation intensity of noise which is superimposed on the signal wave at the laser output. The noise power added to the signal wave over the whole laser cross-section at the output is

$$P_R = \frac{n_2}{n_2 - n_1} hf_{12}\, \delta f(e^{2vL} - 1) \tag{107}$$

To characterize the noise of a four-terminal network in communication systems a noise figure is defined. For this definition a thermal noise source at a reference temperature T_0 is connected to the input of the network. Usually room temperature $T_0 = 290°K$ is chosen as the reference temperature. The noise figure is then given by

$$F = \frac{P_T + P_R}{P_T}$$

where $P_T + P_R$ is the total noise power at the output of the network which results from the thermal noise source at the input and noise sources within the network, and P_T is the noise power at the output which would result from the thermal noise source at the input with the network itself noise free.

In the limit of low frequencies or high temperature the thermal noise power of a source at temperature T is

$$P_0 = kT\delta f \tag{108}$$

In this expression it has been assumed that the equi-partition law (68) holds. This is only true exactly for the limits $f \to 0$ or $T \to \infty$. For finite frequencies and temperatures the general distribution law (70) must be used. Comparing (68) and (70) expression (108) may be generalized by replacing kT with

$$\frac{hf}{e^{\frac{hf}{kT}}-1}$$

Hence the general formula for thermal noise power is

$$P = \frac{hf}{e^{\frac{hf}{kT}}-1}\,\delta f \tag{109}$$

This is the available noise power of a thermal noise source. If the laser in the amplifier arrangement of Fig. 10 is matched at the input and output to the signal wave, all the available noise power will be amplified by the laser. In this case the output power from the thermal noise source at the input is

$$P_T = \frac{hf}{e^{\frac{hf}{kT}}-1}\,\delta f\, e^{2vL}$$

Using this expression and P_R from eqn. (107) the noise figure of the laser amplifier due to spontaneous emission is

$$F = 1 + \frac{n_2}{n_2 - n_1}\left(e^{\frac{hf}{kT}}-1\right)(1-e^{-2vL}) \tag{110}$$

In the ideal situation of a very strong inversion of population density ($n_1 \ll n_2$) and high gain ($e^{2vL} \gg 1$) the noise figure reduces to

$$F = e^{\frac{hf}{kT}} \tag{111}$$

Under these ideal conditions we will also obtain simple expressions for the equivalent noise power P_{RE} and for the equivalent noise temperature T_R at the input. P_{RE} is the available power at the input which will give P_R as output power. Under the present ideal conditions we have

$$P_R = hf_{12}\,\delta f\, e^{2vL}$$

and therefore

$$P_{RE} = hf\,\delta f \tag{112}$$

T_R is the temperature of a thermal noise source which will give P_{RE} as output power. Comparing eqns. (109) and (112) the following equation must be satisfied by T_R:

$$e^{\frac{hf}{kT_R}} - 1 = 1$$

Solving for T_R yields

$$T_R = \frac{hf}{k \ln 2} \tag{113}$$

To illustrate, let the signal to be amplified at optical frequencies have a wavelength $\lambda = 1\ \mu$m. The equivalent noise temperature from (113) for such a wavelength is 21,000°K. The noise of spontaneous emission at optical frequencies is, under the present ideal conditions, much stronger even than thermal noise. The sensitivity of amplifiers for optical frequencies is therefore limited by the noise of spontaneous emission. Because of this limitation amplifiers at optical frequencies can never be as sensitive as microwave amplifiers or even lower frequency amplifiers.

Expressions (112) and (113) for noise due to spontaneous emission under ideal conditions are also obtained for another kind of noise which is of particular significant at optical frequencies. According to eqn. (69)

$$p_{\mathrm{av}} = \frac{1}{e^{\frac{hf}{kT}} - 1} \tag{114}$$

is the average number of photons per mode of oscillation of a thermal radiation field. Under the most favourable conditions for detection and amplification the signal will be in just one mode of the particular arrangement. Due to the quantization of energy in any mode noise will be associated with the signal. Since it is caused by the quantization of energy into photons this noise is called quantum noise. The ratio of the corresponding noise power to the signal power will become larger when the signal is weakened. It increases when the average number of photons in the signal mode decreases. In the case of

$$p_{\text{av}} = 1 \tag{115}$$

the quantum noise power will be just equal to the average signal power.

Comparing (114) and (115) an equivalent noise temperature T_Q may again be defined for a thermal noise source, which in this case has the quantum noise power as available power. To have $p_{\text{av}} = 1$ for the thermal radiation field the following equality must be satisfied by T_Q:

$$e^{\frac{hf}{kT_Q}} - 1 = 1$$

This is the same condition as was obtained for the equivalent noise temperature T_R of spontaneous emission under ideal conditions. T_Q will therefore also be given by eqn. (113).

Under ideal conditions the sensitivity of optical amplifiers will not only be limited by the noise of spontaneous emission, the same limitation is obtained from the quantum nature of all electromagnetic fields. At optical frequencies a single photon has sufficient energy for quantum noise to set the ultimate limit in sensitivity.

2.4. Laser oscillator

To generate oscillations, positive feedback is provided in the laser arrangement. As shown in Fig. 11 the active material is bounded on two opposite sides by mirrors. One or both of these mirrors are made partly transparent. Photons which are spontaneously emitted during transitions from state 2 to state 1 will trigger stimulated emission. Such spontaneously emitted photons travel through the active medium and thus stimulate further transitions of micro-systems from states 2 to 1. The photons from these stimulated transitions add in phase to the stimulating photons and thus amplify the spontaneous radiation. At the end mirror the photons are reflected and travel again through the medium. More stimulated photons are added and the radiation is further amplified. In this process a wave is generated which travels back and forth between both mirrors parallel to the axis of the laser arrangement. Radiation which travels in other directions will leave the laser material through the transparent side walls either at once or after having been

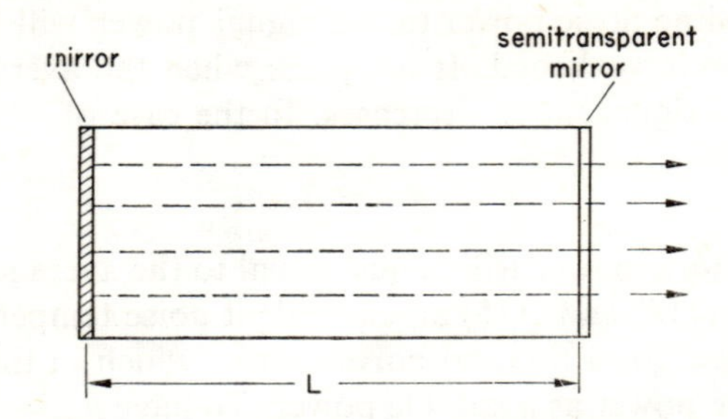

FIG. 11. Optical resonator with active material as regenerative oscillator

reflected only a few times between both mirrors. Such radiation will not generate waves of sufficient amplitude to sustain oscillation. For the waves in the longitudinal direction to lead to self-sustained oscillation, amplification during one transit must be just as large as the attenuation. Attenuation of the paraxial waves in the optical resonator of Fig. 11 has several causes:

Absorption in the laser medium.
Scattering at optical imperfections in the medium.
Transmission and absorption at the end mirrors.
Diffraction at the mirror edges.

All these losses may be included in a total attenuation factor a. As in case of the gain constant v this exponential factor a defines an attenuation constant α when it is divided by the length of the optical resonator

$$\alpha = \frac{a}{L} \tag{116}$$

Alternatively the attenuation in the resonator may be described by a time constant

$$\tau_p = \frac{1}{2\alpha c} = \frac{Q}{\omega} \tag{117}$$

where Q is the quality factor of the respective mode of oscillation in the resonator. When no energy is supplied by stimulated emission or otherwise the respective mode of oscillation would decay exponentially with the time constant τ_p. The subscript p of τ_p indicates that this time constant may also be regarded as the mean lifetime of photons in this mode.

The starting condition for oscillation is

$$v \geqslant \alpha$$

To satisfy this condition the inversion of population densities between the states for stimulated transitions must be

$$\Delta n \geqslant \frac{2c\alpha}{hfB_{12}g(f)}$$

or

$$\Delta n \geqslant \frac{1}{hfB_{12}\tau_p g(f)} \qquad (118)$$

We will replace B_{12} in this expression by the mean lifetime τ_{12} due to spontaneous transitions according to (72) and introduce the spectral mode density n_f according to (67). The threshold of inversion density for oscillation may then be written as

$$\Delta n = \frac{n_f}{g(f)} \frac{\tau_{12}}{\tau_p} = \frac{n_f \Delta f}{g(f)\Delta f} \frac{\tau_{12}}{\tau_p} \qquad (119)$$

The meaning of $n_f \cdot \Delta f$ is the effective number of modes per unit volume within the spectral range Δf of the emission-line. If Δf is the 3-db bandwidth of the emission line then no matter what the particular line shape is the product $g(f)\Delta f$ will be of the order of unity. According to this quite explicit relation there must be a sufficiently greater number of micro-systems in the upper state 2 than in state 1. This number is obtained by multiplying the number of modes $n_f \Delta f$, which are available for spontaneous emission, by the ratio of mean lifetime τ_{12} of micro-systems in the upper state to the mean lifetime τ_p of photons in the modes of the resonator.

When Δn grows from its negative value for thermal equilibrium through zero to positive values during the pumping process the starting condition for oscillation will be satisfied first for that mode which is closest in resonant frequency to the maximum of the emission line. This mode will therefore be the first to start oscillating. When the resonant frequency is right at the centre of the emission line, and when this emission line is represented by a Lorentzian curve, then the starting condition is

$$\Delta n \geqslant \frac{\pi}{2} n_f \Delta f \frac{\tau_{12}}{\tau_p} \qquad (120)$$

When the emission line is a Gaussian curve under the same conditions we will have oscillations when

$$\Delta n \geqslant \sqrt{\frac{\pi}{\ln 2}}\,\frac{n_f}{2}\,\Delta f\,\frac{\tau_{12}}{\tau_p} \tag{121}$$

According to these expressions the threshold value for Δn will be low only when the laser medium has a very sharp emission line and when the optical resonator has a high quality factor.

Of all active micro-systems in the upper state 2 a number $A_{12}\,N_2$ per unit time will undergo spontaneous transitions to state 1. To keep the population inversion ΔN at a constant level the pump source at frequency f_p must, therefore, supply at least the following pumping power

$$P_f = A_{12} h f_p N_2 \tag{122}$$

For this and higher values of pumping power fluorescence will be observed in the laser medium.

In case of the four-level laser we have $N_2 \simeq \Delta N$. The threshold power for fluorescence or for laser action then is

$$P_f = A_{12} h f_p\, \Delta N \tag{123}$$

Using eqn. (121) the threshold power for the four-level laser may also be written as

$$P_f = \sqrt{\frac{\pi}{\ln 2}}\,\frac{h f_p n_f\, \Delta f}{2\tau_p}\,V \tag{124}$$

We note again from this expression that for low threshold values Δf should be small and τ_p large.

2.5. Power and frequency of oscillation

When a laser oscillator is switched on it does not instantaneously deliver power of a constant level at a constant frequency. Instead the oscillation goes through transients before it becomes stationary or reaches a steady state. To describe the transient behaviour of laser oscillators when they are switched on or off differential equations are formulated for the rate of change with time of the population in participating states and for the rate of change of the energy in the electromagnetic field. These differential equations are obtained by considering first the population of states and transitions between states and then taking account of the associated energy changes together with the energy which is stored in the field. Since these differential equations are directly stated for the rate of change with time they are called rate equations. Steady state solutions of these rate equations will give us detailed information about the output power and frequency of the laser oscillation and about saturation characteristics and frequency stability. We will, therefore, first formulate the rate equations in as general a form as is appropriate for this introductory text. We will then limit our discussion to the steady state solutions only in order to discuss output power and frequency of laser oscillations. A more general solution for the transient behaviour will be given in the subsequent section.

The rate equations will be formulated here only for the ideal four-level laser. The lower state 1 of stimulated transitions will be assumed to be always sufficiently depopulated so that $N_2 \gg N_1$. The depletion of state 1 by spontaneous or non-radiative transitions through relaxation is assumed to be so effective that $N_2 \gg N_1$ even when the transition rate from the upper state 2 to 1 is very large. The rate equations may also be formulated for a more realistic four-level laser where the assumption $N_2 \gg N_1$ has been dropped. They may even be formulated for the three-level laser. These more accurate and detailed rate equations and their solutions are too involved to be discussed here. The most important characteristics are included in the model of the ideal four-level laser. It will therefore be sufficient to consider only the simpler rate equations.

We get started more easily by first assuming that only one mode of the laser resonator is excited. Let this mode be a standing wave formed by two uniform plane waves of equal amplitude which travel in opposite directions parallel to the resonator axis. The field energy W in this mode will be measured by the number p of the photons or quanta

$$p = \frac{W}{hf}$$

The time rate of change for the population N_2 of the upper laser state is

$$\frac{dN_2}{dt} = N'_{23} - N_2 B_p p - \frac{N_2}{\tau_{12}} \tag{125}$$

Here N'_{23} indicates the number of micro-systems which make transitions per unit time from the ground state 0 into the upper laser state

2 by way of the pump level 3. The transitions from the ground state to the pumping level 3 are induced by the pumping power, while the transitions from 3 to 2 are non-radiative relaxation processes.

The second term on the right-hand side of eqn. (125) accounts for the radiating transitions from 2 to 1 which are stimulated by the photons p in the oscillating mode. B_p is another form for the Einstein coefficient of stimulated emission which, in the case of a uniform field, is related to the more conventional coefficient B_{12}, which was used in our previous calculations, by

$$B_p = \frac{B_{12}}{V} hfg(f) \tag{126}$$

V is again the volume of the resonator. The last term in eqn. (125) accounts for spontaneous transitions from 2 to 1. The radiation due to these spontaneous transitions triggers the self-sustained oscillations. In the steady state the spontaneous emission causes the spectrum of the output power to have a finite linewidth.

A second differential equation may be formulated for the rate of change with time of photons in the oscillating mode

$$\frac{dp}{dt} = N_2 B_p p - \frac{p}{\tau_p} + \frac{N_2 g(f)}{n_f V \tau_{12}} \tag{127}$$

The first term on the right-hand side accounts for the increase in photons due to stimulated emission. The second term accounts for the reduction of the photon number due to resonator losses. The mean lifetime τ_p of photons in the particular mode is meant to include not only the attenuation within the resonator but also as in eqn. (114) the loss due to output power leaving the resonator through one or both end mirrors.

The last term in eqn. (127) accounts for the increase in the number of photons due to spontaneous emission. With n_f being the spectral mode density and V the resonator volume the product $n_f \cdot V \cdot df$ represents the number of modes in V with resonant frequencies between f and $f+df$. The rate of spontaneous transitions emitting photons within this frequency interval is $N_2 g(f)\, df/\tau_{12}$ when $g(f)$ is the normalized shape of the emission line. The ratio of both these quantities is

$$\frac{N_2 g(f)}{n_f V \tau_{12}}$$

It represents the average number of photons per unit time which are emitted into the resonator mode of frequency f due to spontaneous transitions. According to eqns. (67) and (72) as well as eqn. (126) we have

$$\frac{g(f)}{n_f V \tau_{12}} = B_p \tag{128}$$

Comparing this expression and the above ratio the spontaneous emission into one mode is found to be just as large as the induced emission which, on average, is stimulated by a single photon in the same mode.

In summarizing the two rate equations may be written as follows

$$\frac{dN_2}{dt} = N'_{23} - N_2 B_p p - \frac{N_2}{\tau_{12}} \tag{129}$$

$$\frac{dp}{dt} = N_2 B_p p - \frac{p}{\tau_p} + N_2 B_p \tag{130}$$

They represent two coupled differential equations. The equations are non-linear. Under general conditions only numerical solutions are feasible.

The rate equations are readily solved only for laser oscillations in the steady state. In this case all time derivatives are zero. The rate equations reduce to:

$$N'_{23} - N_2 B_p p - \frac{N_2}{\tau_{12}} = 0 \tag{131}$$

$$N_2 B_p p - \frac{p}{\tau_p} + N_2 B_p = 0 \tag{132}$$

Of all steady state solutions which are possible for laser oscillations let us first examine the limiting case where the self-excitation has just started. From the general results of section 2.4 we have that for this limit the stimulated emission $N_2 B_p p$ just compensates the loss p/τ_p of the resonator. We therefore have

$$N_2 = \frac{1}{B_p \tau_p} \tag{133}$$

On the other hand, in order to satisfy rate equation (132) we must have

$$N_2 = \frac{1}{B_p \tau_p} \frac{p}{1+p} \tag{134}$$

The population inversion N_2 in steady state according to eqn. (134) can never be larger than its threshold value (133) when oscillation has just started. This threshold value is a limit which is approached by N_2 only in the case of very many photons p in the particular mode of oscillation. Once oscillation has started in a particular resonator mode the number of photons actually grows quite rapidly to 10^2 or even 10^5. N_2 is, therefore, nearly equal to its threshold value under normal conditions for oscillation.

In general for any steady state situation the number of photons in a resonator mode may be calculated by solving (131) and (132) for p. To this end we substitute $N_2 B_p p$ from (131) into (132) and solve for p. The result is

$$p = N'_{23}\tau_p - \left(\frac{1}{\tau_{12}} - B_p\right) N_2 \tau_p$$

By further substituting for N_2 from (134) we obtain

$$p = N'_{23}\tau_p - \left(\frac{1}{B_p \tau_{12}} - 1\right) \frac{p}{1+p}$$

If under normal conditions $p \gg 1$, an excellent approximation for the number of photons is

$$p = N'_{23}\tau_p - \frac{1}{B_p \tau_{12}} + 1 \tag{135}$$

Using (128) for the second term on the right-hand side we may also write

$$p = N'_{23}\tau_p - \left(\frac{n_f V}{g(f)} - 1\right) \tag{136}$$

The number of photons in a mode of oscillation is predominantly determined by the product $N'_{23}\tau_p$ of pump rate and mean lifetime of the photons. The term in parentheses on the right-hand side of (136) accounts for the reduction in photons due to spontaneous emission into all other modes within $g(f)$. There are $n_f V/g(f)$ photons being spontaneously emitted per unit time into all modes, less one photon which, on the average, is spontaneously emitted into the oscillating mode.

To calculate the output power of the laser we consider the energy $W = hfp$ which is stored in the oscillating mode. The total power which this mode delivers is

$$P = \frac{W}{\tau_p} = hf\frac{p}{\tau_p}$$

But this power is not all useful output power; part of it is absorbed in the resonator or otherwise lost. If the power which is delivered from the oscillating mode were all useful output power, without any internal losses, and if the resonator in this case had a time constant τ_p' then the output power of the actual resonator with its time constant τ_p is

$$P' = P\frac{\tau_p}{\tau_p'} = hf\frac{p}{\tau_p'}$$

Substituting (136) for the number of photons in the oscillating mode results in

$$P' \simeq \frac{hf}{\tau_p'}\left(N_{23}'\tau_p - \frac{n_f V}{g(f)} + 1\right) \tag{137}$$

To obtain a high output power for a given pumping rate N_{23}' the overall losses of the oscillating mode should be low. The total time constant τ_p will then be large. But not only should these overall losses be low, the useful output power should constitute as large a part of these overall losses as possible.

In addition the product $n_f \cdot V$ should be small in the vicinity of the frequency of oscillation. The resonator should, therefore, have only a low spectral density of modes—that is only a few other modes with resonant frequencies near the frequency of the oscillating mode.

The oscillating mode of the laser resonator is not purely monochromatic. Even though each photon emitted by stimulated transitions has exactly the same frequency and phase as the stimulating photon there are, in addition, spontaneously emitted photons which have random frequency and phase within the resonance width of the oscillating mode. These spontaneous photons cause small but random variations of amplitude and phase of the oscillating mode. The frequency spectrum of the oscillation is therefore not just one line but has finite width.

The 3-db width of this line may be found rather easily from eqn. (132). The 3-db width of the resonance line for a mode in a passive resonator without gain by stimulated emission is

$$\Delta f_p = \frac{1}{2\pi\tau_p} \tag{138}$$

This bandwidth is determined by all the losses in the resonator. In eqn. (132), however, the second term accounts for all losses in the resonator.

Stimulated emission compensates for resonator losses. In eqn. (132) the first term accounts for this compensation of losses due to stimulated emission. According to the first two terms in eqn. (132) the resonance width Δf_p of the passive resonator is reduced to

$$\delta f_p = \frac{1}{2\pi}\left(\frac{1}{\tau_p} - N_2 B_p\right)$$

Substituting from (134) for N_2 we obtain

$$\delta f_p = \frac{\Delta f_p}{1+p} \simeq \frac{\Delta f_p}{p} \tag{139}$$

This then is the resonance width of the oscillating mode. Within the resonance width photons are spontaneously emitted into the mode and thus excite the oscillation. The oscillating mode, therefore, has a frequency spectrum which in form and width corresponds to this resonance line. In the case where $p = 1$ only spontaneous photons are emitted into the mode of oscillation. Under these circumstances we have $\delta f_p = \frac{1}{2}\Delta f_p$. For all $p > 1$ the frequency spectrum of this mode is narrower. Its width will be the smaller the larger the number of photons in the mode or the larger the energy in the particular oscillation.

The output power of this oscillation has the same frequency spectrum. Using eqns. (137) and (138) the half-power bandwidth of the spectrum may be expressed also in terms of this output power P', i.e.

$$\delta f_p \simeq \frac{2\pi h f (\Delta f_p)^2}{P'} \frac{\tau_p}{\tau_p'} \tag{140}$$

Equations (139) and (140) give a lower limit for fluctuations in frequency which will always be present in laser oscillators. Under practical conditions even larger fluctuations due to other, mostly external, causes will be added to the frequency fluctuations due to spontaneous emission.

2.6. Relaxation oscillations

A physical system which is capable of oscillating will only be gradually excited when, upon changing certain parameters (for example the switching on of a bias voltage), it is suddenly brought into the state for which self-oscillation is possible. The oscillations will first grow continuously in amplitude and then gradually approach a steady state. The build-up or transient phenomena of this kind are often called relaxation oscillations.

The laser oscillator is a typical case of such transient phenomena. When pumping radiation is suddenly applied or its resonator characteristics are changed instantaneously it will go through relaxation oscillations.

These transient oscillations are of particular interest when the laser is operated to generate pulses. For pulse generation either the pumping power is applied only for a short duration, or the laser resonator is switched almost instantaneously to a high quality factor.

When pumping power is switched on the pumping rate N'_{23} in the rate equations will suddenly be stepped up to a certain value. Self-oscillation will then start and, during the transient, the number of excited microsystems as well as the number of photons in the particular mode of oscillation will gradually go into the equilibrium state which is given by the stationary form of the rate equations.

The general rate equations are non-linear and solutions are thus difficult to obtain. It will, however, be sufficient for us to consider only an approximate solution. We assume all fluctuations of N_2 and p in time to be small compared with their steady state values. In

$$N_2 = N_0 + N_t \quad \text{and} \quad p = p_0 + p_t \tag{141}$$

let N_0 and p_0 be constant in time, then

$$|N_t| \ll N_0 \quad \text{as well as} \quad |p_t| \ll p_0 \tag{142}$$

Under these assumptions we will not be able to account for the complete transient process. At best we will be able to describe the last phase of it. Nevertheless, the corresponding approximate solution will give us some insight into this process and other relaxation oscillations.

Under condition (142) the non-linear rate equations may be made linear. Substituting for N_2 and p from (141) into (129) and (130) all terms with products of N_t and p_t may be neglected according to (142) since they are small compared with all other terms. The result is

$$\frac{dN_t}{dt} = N'_{23} - N_0 B_p p_0 - \frac{N_0}{\tau_{12}} - B_p[N_0 p_t + p_0 N_t] - \frac{N_t}{\tau_{12}} \tag{143}$$

$$\frac{dp_t}{dt} = N_0 B_p p_0 - \frac{p_0}{\tau_p} + N_0 B_p + B_p[N_0 p_t + (p_0+1)N_t] - \frac{p_t}{\tau_p} \tag{144}$$

The stationary quantities N_0 and p_0 are solutions of the steady state rate equations (131) and (132). Therefore the first three terms on the right-hand sides of (143), as well as (144), add up to zero. Furthermore, under normal conditions of steady state oscillations we have $p_0 \gg 1$, the number of photons in the stationary mode being very large. The time-dependent variations of N and p will therefore obey the following linear differential equations

$$\frac{dN_t}{dt} = -\left[B_p p_0 + \frac{1}{\tau_{12}}\right] N_t - B_p N_0 p_t \tag{145}$$

$$\frac{dp_t}{dt} = B_p p_0 N_t$$

The solutions of this linear and homogenous system are exponential functions for both N_t and p_t

$$N_t = N_{t0} e^{st}, \qquad p_t = p_{t0} e^{st}$$

Upon substituting these exponential functions into (145) a homogeneous system of linear equations for N_{t0} and p_{t0} is obtained. For non-trivial solutions the coefficient determinant of the system must be zero, thus

$$\begin{vmatrix} -s - B_p p_0 - \dfrac{1}{\tau_{12}} & -B_p N_0 \\ B_p p_0 & -s \end{vmatrix} = 0$$

This characteristic equation is a quadratic polynomial in s. Two solutions for s are obtained:

$$s = -\frac{1}{2}\left(B_p p_0 + \frac{1}{\tau_{12}}\right) \pm j\sqrt{B_p \frac{p_0}{\tau_p} - \frac{1}{4}\left(B_p p_0 + \frac{1}{\tau_{12}}\right)^2}$$

In general these values of s are complex. The relaxation is thus a damped oscillation decaying exponentially with a damping coefficient

$$\sigma = \frac{1}{2}\left(B_p p_0 + \frac{1}{\tau_{12}}\right)$$

while oscillating with angular frequency

$$\omega = \sqrt{B_p \frac{p_0}{\tau_p} - \frac{1}{4}\left(B_p p_0 + \frac{1}{\tau_{12}}\right)^2}$$

For the stationary number p_0 of photons we may substitute from (135). If again we assume $p_0 \gg 1$ then the damping coefficient and characteristic frequency of the relaxation oscillation reduce to

$$\sigma = \tfrac{1}{2} B_p N'_{23} \tau_p \tag{146}$$

$$\omega = \sqrt{B_p N'_{23} - \frac{1}{\tau_p \tau_{12}} - \tfrac{1}{4} B_p^2 N'^2_{23} \tau_p^2} \tag{147}$$

In the case of

$$B_p N'_{23} < \tfrac{1}{4} B_p^2 N'^2_{23} \tau_p^2 + \frac{1}{\tau_p \tau_{12}}$$

the relaxation decays aperiodically. For this aperiodic situation the damping coefficient is

$$\sigma = \tfrac{1}{2} B_p N'_{23} \tau_p \mp \sqrt{\tfrac{1}{4} B_p^2 N'^2_{23} \tau_p^2 + \frac{1}{\tau_p \tau_{12}} - B_p N'_{23}}$$

A more detailed examination of these approximate solutions has revealed that the agreement between the present linear approximation and a more accurate solution of the non-linear rate equations is good only when $\omega > \sigma$. Also only in the case where pronounced relaxation oscillations occur will the present approximations for ω and σ correspond to the actual characteristics of the transients.

We will therefore consider only the range $\omega > \sigma$ when we have pronounced relaxation oscillation. Furthermore, we will assume that the frequencies of only a few modes of the resonator are close to the self-excited modes and that only a few photons are emitted into all other modes. In eqn. (135) this means that

$$N'_{23} \tau_p \gg \frac{1}{B_p \tau_{12}}$$

Under these conditions the approximations for the damping coefficient and frequency of the relaxation oscillation now take the form

$$\sigma = \tfrac{1}{2} B_p N'_{23} \tau_p$$
$$\omega = \sqrt{B_p N'_{23}} \tag{148}$$

If the pumping radiation is low or just barely above the threshold of fluorescence the relaxation oscillations will be only weakly attenuated. The damping will be more pronounced when more pump power is applied. The frequency of the relaxation oscillation will also increase with pumping power. For very intense pumping radiation the relaxation will eventually change into an aperiodic transient.

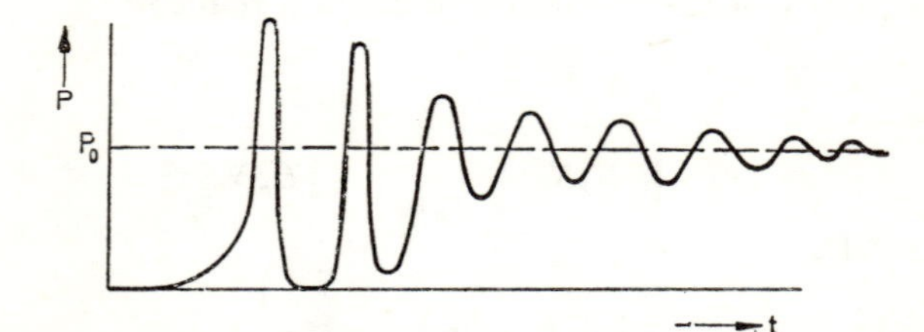

FIG. 12. Relaxation oscillations during the initial transient of a solid-state laser

The initial transient for a solid-state laser with $CaF_2 : U^{+3}$ is shown in Fig. 12. Such a laser will be described in more detail in a subsequent section. When this laser is first switched on there are pronounced spikes of emission. Of course such spikes cannot be accounted for by a linear approximation. During the tail of the transient process when the variations become smaller the damped oscillation occurring in the relaxation can be recognized.

2.7. Multimode oscillations

During the preceding discussion of the steady-state oscillations in lasers we have always assumed that only one particular mode of the optical resonator is actually excited. In most practical situations this assumption is really not justified. On the one hand the emission lines of most laser materials are sufficiently wide and on the other hand the spectral density of modes is so high that the threshold of oscillation will be reached for many modes at the same time. Oscillations will then be excited in many modes simultaneously.

An optical resonator, as shown in Fig. 11, but of unlimited cross-section has modes with a uniform field structure in the cross-sectional planes. All these modes are standing waves which are formed by two uniform plane waves travelling in opposite directions exactly parallel to the resonator axis.

For two such modes which are adjacent in frequency the number of standing half-waves differs by only one. If n is the number of half-wave lengths and L the length of the resonator then $f_1 = \frac{nc}{2L}$ is the frequency of oscillation of this particular mode. $f_2 = \frac{(n+1)c}{2L}$ is the frequency of the corresponding mode next higher in frequency. Their frequency difference is

$$f_2 - f_1 = \frac{c}{2L}$$

In most cases the emission lines of laser materials are much wider than this frequency difference. In addition in between f_1 and f_2 there are frequencies of other modes which are formed by uniform plane waves travelling in nearly opposite directions at a small angle with respect to the axis of the resonator. Their field distribution is not uniform over the cross-section of the resonator. Some of these modes may also be excited. The complete mode spectrum of practical resonators will be discussed in a subsequent chapter.

The number of modes and their amplitudes will not only be determined by their spectral density and their quality factor, but also by the nature of the line broadening of the emission line, whether it is homogeneous or inhomogeneous. In the case of a homogeneously broadened line a photon in any mode is capable of inducing a transition in any one of the active micro-systems, provided its frequency is within the emission line. The situation is different for an inhomogeneously broadened line. Here photons in a particular mode will be able to induce transitions only in those micro-systems where the transition frequencies differ from the oscillating frequency by less than the natural line width of the transition.

Following the self-excitation of certain modes a steady state is established when the amplification of the laser material reaches into saturation.

The inversion of states will become stationary at their threshold values given by eqn. (134). In the case of homogeneous line broadening the gain will be saturated over the full width of the emission line through any one particular mode of oscillation (Fig. 13). The saturation of the emission line is homogeneous. In the case of inhomogeneous line broadening the saturation of the emission line will also be inhomogeneous. As shown in Fig. 14 the inhomogeneously broadened line will be

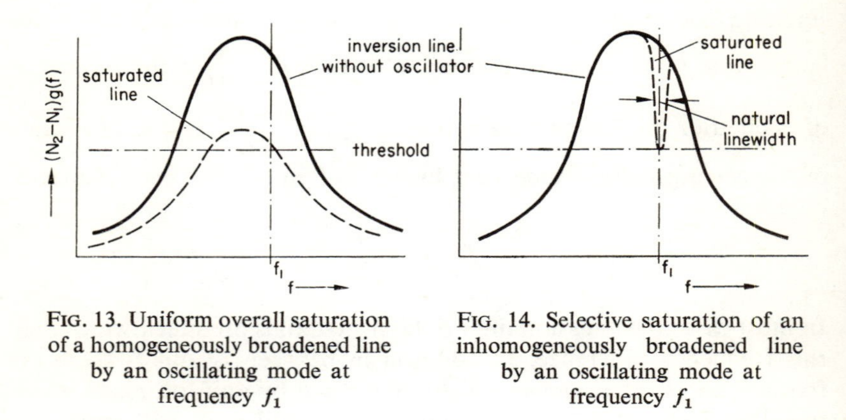

FIG. 13. Uniform overall saturation of a homogeneously broadened line by an oscillating mode at frequency f_1

FIG. 14. Selective saturation of an inhomogeneously broadened line by an oscillating mode at frequency f_1

selectively saturated by individual modes of oscillation. Each mode when it is excited "burns a hole" in the inhomogeneously broadened line.

We will first discuss the simultaneous excitation of many modes in the case of a homogeneously broadened line. This discussion will again be based on the model of the ideal four-level laser. The rate equation (132) still holds for the steady state of each of the many modes which are oscillating simultaneously. The number of photons in a particular mode ν is:

$$p_\nu = \frac{N_2 B_{p\nu} \tau_{p\nu}}{1 - N_2 B_{p\nu} \tau_{p\nu}}$$

The optical resonator of volume V has a number

$$n_p = n_f V \Delta f$$

of modes within the width of the emission line. The Einstein coefficient of eqn. (128) for mode ν may therefore be written as

$$B_{p\nu} = \frac{g(f_\nu)\Delta f}{n_p \tau_{12}}$$

Using this expression we obtain the following equation for the steady state number of photons in the mode ν:

$$p_\nu = \frac{N_2 \tau_{p\nu}}{n_p \tau_{12}} \frac{g(f_\nu)\Delta f}{1 - \frac{N_2 \tau_{p\nu}}{n_p \tau_{12}} g(f)\,\Delta f}$$

Homogeneously broadened line shapes normally follow the Lorentz curve

$$g(f)\,\Delta f = \frac{2}{\pi} \frac{1}{1 + \left(2 \frac{f - f_0}{\Delta f}\right)^2}$$

With such a line shape the steady state number of photons is

$$p_\nu = \frac{2N_2 \tau_{p\nu}}{\pi n_p \tau_{12}} \frac{1}{\left(2 \frac{f_\nu - f_0}{\Delta f}\right)^2 + \left(1 - \frac{2N_2 \tau_{p\nu}}{\pi n_p \tau_{12}}\right)} \tag{149}$$

Let us assume here that all modes have the same loss, then all the photon lifetimes will be identical,

$$\tau_{p\nu} = \tau_p = \text{const.} \tag{150}$$

independently of the particular frequency f_ν of oscillation. Under these circumstances p_ν as a function of f_ν is again a Lorentz curve. As shown in Fig. 15 this Lorentz curve forms the envelope for all values p_ν. Its 3-db bandwidth is

$$\delta f = \Delta f \sqrt{1 - \frac{2N_2 \tau_p}{\pi n_p \tau_{12}}} \tag{151}$$

The envelope is always contracted compared with the original emission line. The envelope will become even narrower as the applied pump power is increased.

A more explicit and useful formula for p_ν and the bandwith δf of its envelope will be obtained when the population inversion N_2 is expressed

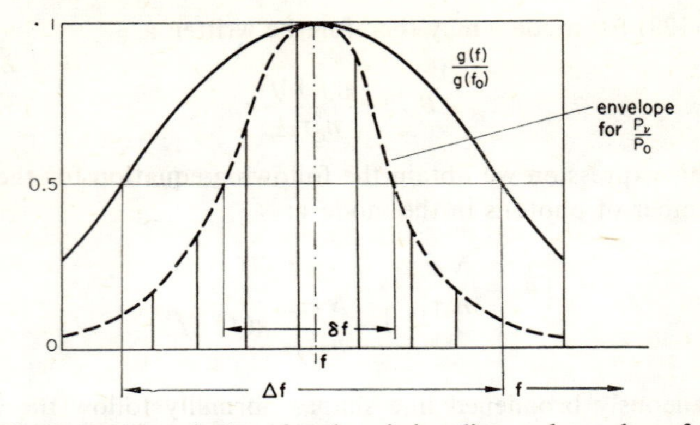

FIG. 15. Homogeneously broadened emission line and envelope for the energy of individual modes of oscillation

in terms of the number p_0 of photons which would be in a mode at centre frequency $f_\nu = f_0$ at the steady state. Using rate equation (132) this expression is

$$N_2 = \frac{1}{B_{p0}\tau_{p0}} \frac{p_0}{1+p_0}$$

Substituting into (149) the result for modes with arbitrary $\tau_{p\nu}$ is

$$p_\nu = \frac{p_0}{1+p_0} \frac{\tau_{p\nu}}{\tau_{p0}} \frac{1}{\left(2\frac{f_\nu - f_0}{\Delta f}\right)^2 + 1 - \frac{p_0}{1+p_0} \frac{\tau_{p\nu}}{\tau_{p0}}} \tag{152}$$

Taking the restriction of equal loss in all modes according to (150) the 3-db bandwidth of the envelope for all p_ν is then

$$\delta f = \frac{\Delta f}{\sqrt{1+p_0}} \tag{153}$$

The mode 0 at centre frequency f_0 will be excited most strongly. At the steady state the number p_0 of photons is therefore normally quite large. As a consequence the envelope which describes the energy distribution among the many simultaneously oscillating modes is much narrower than the original emission line. If we pump harder p_0 will increase further and the envelope will then be contracted even more. Just above threshold many modes within the emission line will be excited, even if only weakly. If however the pumping radiation is intensified then the energy of induced emission will be increasingly concentrated within the modes at the centre of the emission line. Furthermore, in this process the energy as a whole is also increased.

In the case of inhomogeneous line broadening the analysis will be facilitated if we assume that between adjacent oscillating modes we have a separation in frequency which is always greater than the natural line width $(\Delta f)_0 = 1/2\pi\tau_{12}$ of the transition. In general there will be other modes in between these oscillating modes which are not excited however because of high losses. Each oscillating mode will then burn its own hole in the inhomogeneously broadened line and will be independent of all other modes which are oscillating simultaneously.

To obtain information about the steady state power of each oscillating mode we make use of eqn. (136). The total pumping rate N'_{23} generates a population inversion which is uniformly distributed over all micro-systems. Of these micro-systems a fraction $g(f)\,df$ have transition frequencies between f and $f+df$.

The pumping rate which is associated with this fraction is therefore $N'_{23}g(f)\,df$. Let Δf_ν be the frequency separation between simultaneously oscillating modes and let all these modes be equally spaced in frequency by Δf_ν. Furthermore assume the loss factors of all these modes to be equal, according to

$$\tau_{p\nu} = \tau_p$$

All these conditions are satisfied by modes with uniform cross-sectional fields in a parallel plane resonator of unlimited cross-section.

Under these conditions the pumping rate associated with the frequency range Δf_ν of one particular mode is $N'_{23}g(f_\nu)\,\Delta f_\nu$. Using eqn. (136) the number of photons in such a mode is given by

$$p_\nu \simeq N'_{23}\tau_p g(f_\nu)\Delta f_\nu - \frac{n_f V}{g(f_\nu)}\,\frac{\Delta f_\nu}{\Delta f} + 1 \tag{154}$$

The factor $\Delta f_\nu/\Delta f$ in the second term on the right-hand side of (154) accounts for the reduction in spontaneous emission compared with the corresponding term in eqn. (136) for spontaneous emission in single mode operation. Associated with the pump rate $N'_{23}g(f_\nu)\,\Delta f_\nu$ we now have a spontaneous emission of photons within the frequency interval Δf_ν only.

In eqn. (154) the first term on the right-hand side is dominant. Its frequency dependence is given by $g(f_\nu)$ and corresponds to the original emission line. The distribution of energy among the simultaneously oscillating modes therefore corresponds to this emission line. The envelope of this distribution is the inhomogeneously broadened line. If we pump harder N'_{23} will increase and with it the total energy in all modes. The relative distribution of this total energy among the oscillating modes will remain unchanged, except that additional modes will be excited as their threshold is exceeded with increasing pump rate. The situation is illustrated in the diagrams of Fig. 16.

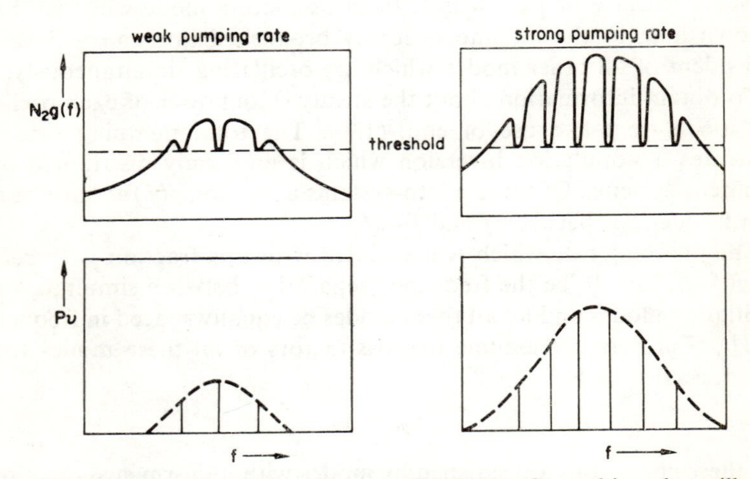

FIG. 16. Selective saturation and energy distribution for multi-mode oscillation with an inhomogeneously broadened line

2.8. Optical resonators

To generate oscillations at optical frequencies the active laser medium is placed in a resonator which, as shown in Fig. 11, consists of two mirrors of high reflectivity. The operation of such a so-called Fabry–Perot resonator will become clear if we assume an electromagnetic wave within the resonator which is repeatedly reflected by the mirrors and steadily amplified while travelling back and forth. To attain self-excitation the wave, after one cycle with reflections at each mirror, must add in phase so that its amplitude will increase through constructive interference. For this phase condition the length of the resonator must be an integral number of half wavelengths. Furthermore, oscillations will only be excited for those waves which travel parallel to the axis of the resonator. Other waves which propagate in off-axis directions will leave through the sides of the resonator after a few reflections. For this reason the plane mirrors of the resonator must be exactly parallel.

A resonator with suitably curved mirrors as in Fig. 17 will have much more favourable characteristics than the plane-parallel Fabry–Perot resonator. In the resonator of Fig. 17 waves which do not travel exactly

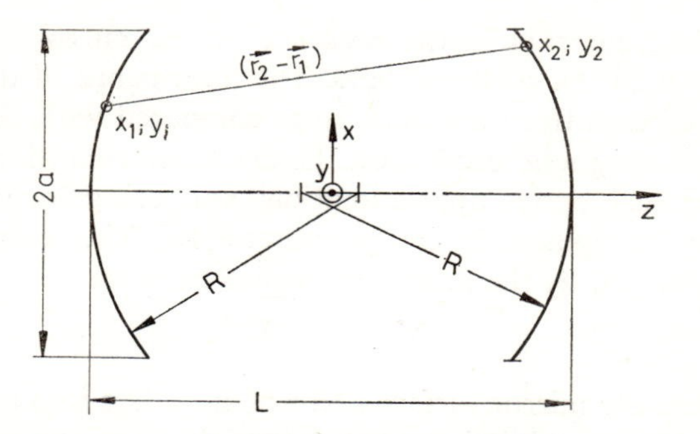

FIG. 17. Optical resonator with spherically curved mirrors

parallel to the axis will nevertheless nearly always be reflected back into the resonator. Likewise the curved mirrors need not be aligned exactly to prevent a loss of waves by radiation through the sides. It is apparent from these considerations that a confocal spacing is the most favourable arrangement of curved mirrors for an optical resonator. In this case the centre of curvature of one mirror is on the surface of the other mirror. Any ray of light through such a centre point is always reflected back into it regardless of its initial direction.

By analogy to the cavity resonators for acoustic waves or microwaves the optical resonators also have natural modes of oscillation. Once excited such a mode would continue to oscillate indefinitely if no energy were lost due to absorption at the mirrors and radiation through the sides. These energy losses will cause the oscillation to decay. Within the laser oscillator the loss in energy is continuously compensated by induced emission and the oscillation is continuous and undamped.

To understand laser operation completely and to design a laser for optimum performance we must have a detailed knowledge of the optical resonator modes. The analysis may be started conveniently from the basic idea that waves are reflected back and forth and always repeat in phase and amplitude distribution after one cycle. In Fig. 17 a wave is assumed which starts from the right-hand mirror and is incident on the left mirror. Let $\mathbf{E}_i^{(1)}$ be the electric field of this incident wave. According to an induction theorem for electromagnetic diffraction [16] the field which is scattered back from the mirror may be calculated from the magnetic surface currents

$$\mathbf{M} = \mathbf{n} \times \mathbf{E}_i^{(1)}$$

These surface currents in the presence of the mirror are equivalent sources for the back scattered field. The dimensions of the mirror and its radius of curvature are very large compared with the wavelength of the electromagnetic oscillation. Under these conditions of the "so-called" physical optics approximation, and using image theory, the mirror may be replaced by image sources [16]. The scattered field will now be generated by the magnetic surface currents

$$\mathbf{M} = 2\mathbf{n} \times \mathbf{E}_i^{(1)}$$

which replace the mirror surface and which radiate into free space. The resulting free space electric vector potential is

$$\mathbf{F} = \iint_{S_1} \frac{\mathbf{n} \times \mathbf{E}_i^{(1)} e^{-jk|\mathbf{r}-\mathbf{r}_1|}}{2\pi|\mathbf{r}-\mathbf{r}_1|} dS_1 \tag{155}$$

where the integration extends over the surface S_1 of the left-hand mirror, $\mathbf{r}_1$ is the position vector of a source point and $\mathbf{r}$ the position vector where the field is evaluated. The scattered field is obtained from (156)

$$\mathbf{E}_i^{(2)} = -\nabla \times \mathbf{F} \tag{156}$$

which, in turn, is incident on the right-hand mirror. For any point with position vector $\mathbf{r}_2$ on the right-hand mirror this incident field is

$$E_i^{(2)} = -\nabla \times \iint_{S_1} \frac{n \times \mathbf{E}_i^{(1)} e^{-jk|\mathbf{r}_2-\mathbf{r}_1|}}{2\pi|\mathbf{r}_2-\mathbf{r}_1|} dS_1 \tag{157}$$

By postulating that the field distribution $\mathbf{E}_i^{(2)}$ of the incident wave on the right-hand mirror is, except for a constant factor, equal to the field distribution $\mathbf{E}_i^{(1)}$ of the incident wave on the left-hand mirror, i.e.

$$\varkappa \mathbf{E}_i^{(2)} = \mathbf{E}_i^{(1)} \tag{158}$$

a homogeneous integral equation is obtained from (157). Solutions of this integral equation describe the natural modes of the resonator. From the corresponding eigenvalues for $\varkappa$ the frequency of these modes and their loss factors due to refraction at the mirror edges may be determined.

To facilitate the discussion of the integral equation we will assume that, in addition to $a \gg \lambda$, the resonator is a thin one, i.e. $L \gg a$. Then in the denominator of the integral the distance between points on the respective mirrors may be approximated by $|\mathbf{r}_2 - \mathbf{r}_1| \simeq L$. However, the exponent in the numerator requires a better approximation. It is not even sufficient to use the Fraunhofer diffraction approximation which, in Cartesian coordinates, would be

$$|\mathbf{r}_2 - \mathbf{r}_1| = L - \frac{1}{L}(x_1 x_2 + y_1 y_2)$$

Rather we will have to take the more accurate form from Fresnel diffraction, which considers all terms up to second order in x and y:

$$|\mathbf{r}_2 - \mathbf{r}_1| = L - \frac{1}{L}\left[x_1 x_2 + y_1 y_2 - \frac{1}{2}\left(1 - \frac{L}{R}\right)(x_1^2 + x_2^2 + y_1^2 + y_2^2)\right]$$

For evaluating the field using $\mathbf{E} = -\nabla \times \mathbf{F}$ it suffices to use radiation field formulae [16]. Assuming that the incident field $\mathbf{E}_i^{(1)}$ is linearly polarized in the x-direction then, since all practical mirror resonators have only slight curvature, the y-component of $\mathbf{M}$ will be the only significant one. $\mathbf{F}$ also will then have only a y-component which, in turn, will yield a field $\mathbf{E}_i^{(2)}$ which is also simply x-directed:

$$E_x^{(2)} = -jkF_y$$

Taking account of all these approximations eqns. (157) and (158) reduce to the following scalar integral equation:

$$E(x_2, y_2) = \varkappa \frac{jk}{2\pi L} e^{-jkL} \iint E(x_1, y_1) e^{j\frac{k}{L}\left[x_1 x_2 + y_1 y_2 - \frac{1}{2}\left(1 - \frac{L}{R}\right)(x_1^2 + y_1^2 + x_2^2 + y_2^2)\right]} dx_1 dy_1 \tag{159}$$

In the case of resonators with rectangular mirrors the limits of integration in Cartesian coordinates are constant and independent quantities, say $x = \pm a_x$ and $y = \pm a_y$. Under these conditions a product solution

$$E(x, y) = u(x)\cdot v(y)$$

will separate the integral equation (159) into two independent equations

$$u(x_2) = \frac{\varkappa_x}{\sqrt{\lambda L}} \int_{-a_x}^{a_x} K(x_1, x_2) u(x_1)\, dx_1$$

$$v(y_2) = \frac{\varkappa_y}{\sqrt{\lambda L}} \int_{-a_y}^{a_y} K(y_1, y_2) v(y_2)\, dy_1 \tag{160}$$

In these equations we have included je^{-jkL} with the coefficients $\varkappa_x$ and $\varkappa_y$; we have therefore

$$j\varkappa = \varkappa_x \varkappa_y e^{jkL}$$

Furthermore the kernel of each integral equation has been represented by

$$K(x_1, x_2) = e^{j\frac{k}{L}\left[x_1 x_2 - \frac{1}{2}\left(1-\frac{L}{R}\right)(x_1^2+x_2^2)\right]} \tag{161}$$

Only two special cases are known for which solutions of the integral equation (160) can be expressed in closed form through available functions. One case is with mirrors of unlimited dimensions (a_x and $a_y \to \infty$), the other case is a confocal arrangement of mirrors with $L = R$.

First we will discuss resonators with infinitely large mirrors. In this case, where $a_x \to \infty$ and $a_y \to \infty$, and putting $L = 2z$, solutions of (160) are [17]:

$$E_{mn} = E_0 \frac{w_0}{w} H_m\left(\sqrt{2}\,\frac{x}{w}\right) H_n\left(\sqrt{2}\,\frac{y}{w}\right) e^{-\frac{x^2+y^2}{w^2}} \tag{162}$$

where

$$w = w_0 \sqrt{1+\left(\frac{2z}{b}\right)^2}$$

and

$$b = w_0^2 k = \sqrt{L(2R-L)} \tag{163}$$

b is called the confocal radius of curvature because $R = b$ for $L = R$. The functions $H_m(\sqrt{2}\,x/w)$ and $H_n(\sqrt{2}\,y/w)$ in (162) are Hermite polynomials of order m and n. They are defined as follows [18]:

$$H_n(x) = (-1)^n e^{x^2} \frac{d^n}{dx^n} e^{-x^2} = n! \sum_{\nu=0}^{\left[\frac{n}{2}\right]} \frac{(-1)^\nu (2x)^{n-2\nu}}{\nu!\,(n-2\nu)!} \tag{164}$$

where $\left[\frac{n}{2}\right] = \frac{n}{2}$ for n even and $\left[\frac{n}{2}\right] = \frac{n-1}{2}$ for n odd. Some of the lower order Hermite polynomials take the form

$$H_0(x) = 1 \qquad H_1(x) = 2x$$
$$H_2(x) = 4x^2-2 \qquad H_3(x) = 8x^3-12x$$

Equation (162) describes the transverse field distribution of the normal modes not only on the mirror faces but everywhere between the mirrors. For this generalization of (162) the quantity z must be interpreted not only as half the resonator length but more generally as the longitudinal coordinate. As is indicated in Fig. 17 the origin of this coordinate z is in the plane of symmetry of the resonator. From this generalization of (162) we obtain as an example the field distribution over the transverse plane at $z = 0$:

$$E_{mn} = E_0 H_m\left(\sqrt{2}\frac{x}{w_0}\right) H_n\left(\sqrt{2}\frac{y}{w_0}\right) e^{-\frac{x^2+y^2}{w_0^2}} \qquad (165)$$

This field distribution is shown in Fig. 18 as a function of one of the transverse coordinates for some of the lower order modes. It is common practice to designate the modes by TEM indicating that their

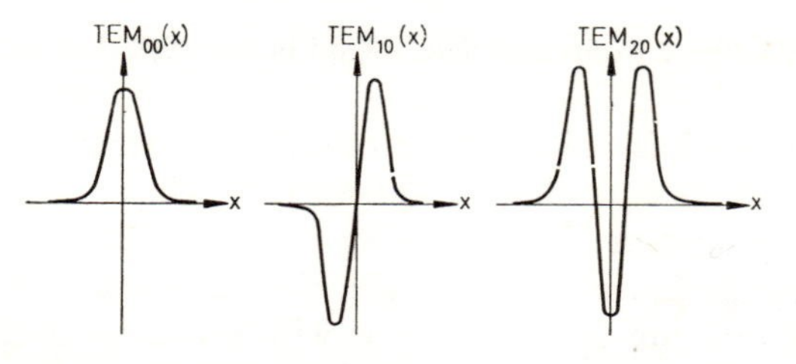

FIG. 18. Transverse field distribution of low order modes in the optical resonator with mirrors of infinite aperture

field distribution is predominantly transverse electromagnetic with respect to their phase fronts. The index numbers of this designation correspond to the transverse orders m and n in eqns. (162) and (165).

The mirrors are surfaces of constant phase for all the modes. Within the resonator, surfaces of constant phase are still spherically curved. Their radius of curvature R depends on the axial position z:

$$R = \frac{b^2+(2z)^2}{4z}$$

The radius of curvature in this equation is positive if, according to Fig. 19, the centre of curvature is to the left of the phase surface. Otherwise it is negative. In the centre of the resonator at $z = 0$ the phase surface is plane. It coincides with the centre plane of symmetry.

The parameter w describes the field behaviour transverse to the axis in cross-sectional planes. The fields decay exponentially in the x- or y-direction at a rate which is given by w. According to (163), w is of the same order of magnitude as $\sqrt{L\lambda}$. Our present assumption of infi-

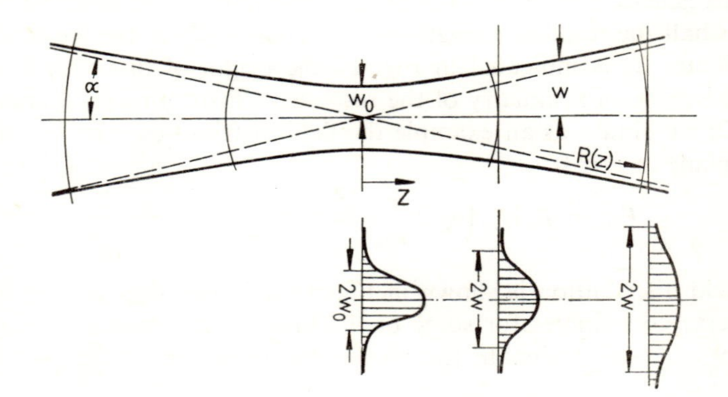

FIG. 19. Phase fronts and field spread in the optical resonator

nitely large mirrors is justified when w or $\sqrt{L\lambda}$ is in some measure smaller than the dimensions a_x and a_y of the mirrors. Under these conditions the low order modes at least will have very small field amplitudes at the mirror boundaries. If, for example, the resonator is 1 m long and oscillates at $\lambda = 1$ μm the quantity w will be of the order of 1 mm. In this case we can neglect the diffraction at their edges entirely when the mirrors have the dimensions of a few millimetres.

At $z = 0$ we have $w = w_0$ and it is here in the centre of the resonator that the transverse field extension of all modes is the smallest. Away from the centre plane it increases in both directions. The exponential in (162) will be a constant for

$$x^2+y^2 = \text{const. } w^2 \tag{166}$$

With w^2 from (163) the transverse field extension therefore grows hyperbolically with z. On the particular hyperboloid

$$x^2+y^2 = w^2 \tag{167}$$

the exponential has decayed to $1/e$ of its value on the resonator axis. Also the field amplitude of the lowest order mode has decayed to $1/e$ here. The quantity w is therefore commonly called the *spot size* of the lowest order mode or of the resonator *per se*.

For very large values of z the hyperboloid of (167) becomes asymptotically conical according to

$$x^2+y^2 = \left(\frac{\lambda}{\pi w_0}\right)^2 z^2$$

The corresponding cone angle is

$$\alpha = \arctan \frac{\lambda}{\pi w_0} \tag{168}$$

A radiating aperture at $z = 0$ with a source distribution corresponding to one of the modes in (165) would have a radiation characteristic with this cone angle α. Furthermore the field of this source at any distance would have spherical phase surfaces and the field distribution over any such phase surface would correspond exactly to the particular mode in the radiating aperture at $z = 0$. Instead of such a radiating field distribution in a plane at $z = 0$ we could also start from the corresponding curved distribution at $z = -L_1/2$ with a curvature radius $-R_1$. This source distribution generates waves with spherical phase fronts travelling in the z-direction. Along the z-axis the position of the phase fronts will be given by [17]:

$$\phi(z) = -k\left(\frac{L_1}{2}+z\right)+(m+n+1)\arccos\sqrt{\left(1-\frac{L_1+2z}{2R_1}\right)\left(1-\frac{L_1+2z}{R(z)}\right)} \tag{169}$$

Within the optical resonator such a wave will be reflected by the mirrors and, by travelling back and forth, will form a standing wave. Whenever the phase of (169) is an integral multiple of π this standing wave fits into the resonator and constitutes a normal mode of oscillation.

According to this general approach the field distributions of (162) will not form just the symmetrical resonator modes found as solutions of (160). In the more general representation of Fig. 19 we may also place spherical mirrors of suitable curvature at any position along the z-axis. Provided these mirrors coincide with the spherical phase fronts and are so spaced that the phase in (169) is an integral multiple of π the field distributions of (162) will still be modes of the new resonator.

By the same idea, for any two different spherical mirrors spaced a certain distance, a plane $z = 0$ may be found and a minimum spot size w_0 be obtained so that the spherical phase fronts associated with the field distributions in (162) coincide with these mirrors. If one of the mirrors is curved with radius R_1 and the other with radius R_2 and if they are spaced by L, then from (169) the total phase shift of the corresponding fields between these mirrors is

$$\phi = kL-(m+n+1) \text{ arc cos} \sqrt{\left(1-\frac{L}{R_1}\right)\left(1-\frac{L}{R_2}\right)}$$

To obtain real values for the phase shifts along z from this expression, the mirror spacing and curvatures must satisfy the following condition

$$0 \leqslant \left(1-\frac{L}{R_1}\right)\left(1-\frac{L}{R_2}\right) \leqslant 1 \tag{170}$$

Only under this condition will there exist real oscillating frequencies and lossless modes. Otherwise the arrangement will not function as a resonator. Figure 20 is a graphical presentation of (170). It indicates all the mirror geometries which will form lossless resonators and have stable modes of oscillation. From this stability diagram either the centre of curvature of one mirror or the mirror itself, but not both, must be located on the axis between the other mirror and its centre of curvature. The Fabry–Perot resonator with plane mirrors and an arrangement with concentric mirrors are the limiting cases.

For a given length of the resonator a mirror curvature may be determined which will have a minimum spot size w. From (163) this minimum spot size w is found for the confocal arrangement of mirrors with $R = b$. The spot size on both mirrors is in this case

$$w = \sqrt{2}\, w_0 = \sqrt{\frac{b\lambda}{\pi}} \tag{171}$$

Sometimes, however, a larger beam diameter is required than would be obtained in the confocal resonator. A typical example is a laser for high power where a large cross-section of the active medium must be utilized. For a large spot size the mirror curvature must be small according to (163) so that the Fabry–Perot resonator with plane mirrors

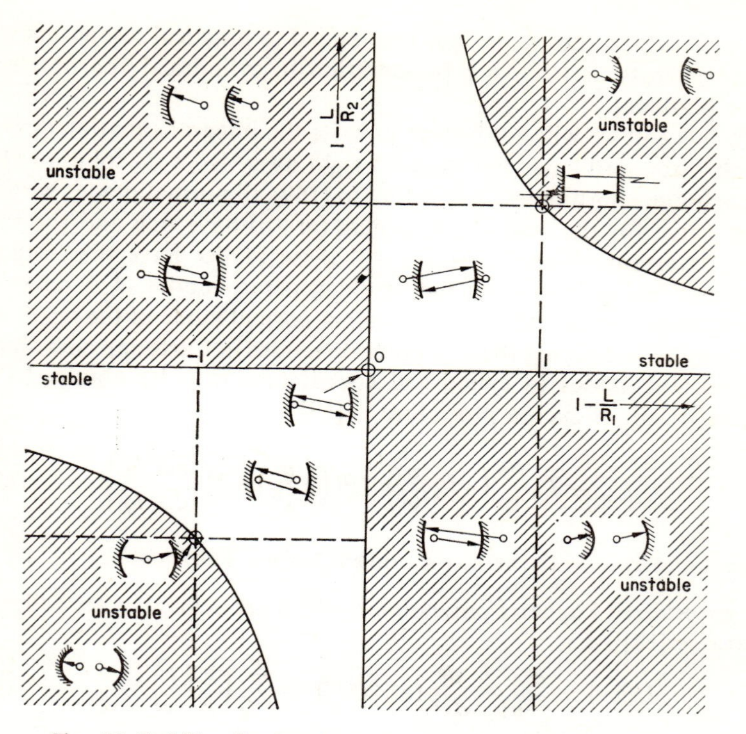

FIG. 20. Stability diagram for modes in optical resonators [23]

is approached. In the case of $|R| \gg L$ the spot size is approximately

$$w = \left(\frac{\lambda z}{\pi}\right)^{\frac{1}{2}} \left(\frac{2R}{L}\right)^{\frac{1}{4}} \tag{172}$$

For plane mirrors without any diffraction at the mirror edges the spot size becomes infinite.

The integral equation (159) for modes in optical resonators and the solutions of this equation have been written in Cartesian coordinates. For rectangular mirrors, using Cartesian coordinates, separation into two simpler integral equations was possible. If the mirrors are round cylindrical coordinates (ϱ, φ, z) must be used. It is true that up to now we have discussed solutions only for mirrors of infinite dimensions. In that case the choice between the two coordinate systems is irrelevant.

When, however, these solutions are applied to practical resonators with round mirrors or irises a representation in cylindrical coordinates is more pertinent.

To this end we will not bother to write the integral equation (159) in cylindrical coordinates but proceed right away from its solution (162) in Cartesian coordinates. According to (169) all normal modes with the same value for the sum of the orders $(m+n)$ have one and the same phase constant along the z-axis. All these modes have therefore the same oscillating frequency; they are degenerate modes. Any combination of such degenerate modes is again a normal mode of the resonator. By suitably combining these degenerate modes in Cartesian coordinates the normal modes in cylindrical coordinates may be obtained. The general result of such a procedure is [19]

$$E_{lp} = E_0 \frac{w_0}{w} \left(\sqrt{2}\,\frac{\varrho}{w}\right)^l L_p^{(l)}\left(2\,\frac{\varrho^2}{w^2}\right) e^{-\frac{\varrho^2}{w^2}} \begin{Bmatrix} \sin l\varphi \\ \cos l\varphi \end{Bmatrix} \tag{173}$$

Here w and w_0 have the same meaning as before. The functions $L_p^{(l)}\left(2\,\frac{\varrho^2}{w^2}\right)$ are generalized Laguerre polynomials which are defined by the following relations [18]:

$$L_p^{(l)}(x) = \frac{e^x x^{-l}}{p!}\,\frac{d^p}{dx^p}(e^{-x}x^{p+l}) \equiv \sum_{\nu=0}^{p} \binom{p+l}{p-\nu} \frac{(-x)^\nu}{\nu!} \tag{174}$$

The three lowest order Laguerre polynomials are

$$L_0^{(l)} = 1, \qquad L_1^{(l)} = l+1-x$$
$$L_2^{(l)} = \tfrac{1}{2}(l+1)(l+2)-(l+2)x+\tfrac{1}{2}x^2$$

The lowest order mode with $p = l = 0$ is the same in both coordinate systems.

In the optical resonator with perfect mirrors of infinite dimensions all modes are loss-less and all eigenvalues of equation (155) have the same magnitude

$$|\varkappa| = 1.$$

Practical resonators have losses which among other causes are due to only partial reflection at the actual mirrors and diffraction at the mirror edges. Uniform reflection losses will leave the field distribution of modes unchanged as in the ideal resonator. Mode losses due to partial reflections are therefore readily determined.

Diffraction at the mirror edges will change the field distribution of the modes. This change is most pronounced in the case of plane mirrors where the spot size for unlimited mirrors is also unlimited. With finite mirrors only modes of finite size can be supported. Diffraction losses in general can only be determined by numerically solving the integral equation with the particular limits of integration.

There is, however, one exception, the confocal resonator. In this case with $L = R$ and rectangular mirrors the following integral equation must be solved

$$u(x_2) = \frac{\varkappa_x}{\sqrt{\lambda L}} \int_{-a}^{a} u(x_1) e^{j \frac{k}{L} x_1 x_2} dx_1$$

The right-hand side is a limited Fourier transform with $x_1 = \pm a$ as the limits. Solutions of this integral equation are [20]

$$u_m = S_m\left(\frac{ka^2}{L}, \frac{x}{\alpha}\right) \tag{175}$$

with eigenvalues

$$\frac{1}{\varkappa_m} = e^{-jkL} \sqrt{\frac{ja^2}{L\lambda}}\, 2j^m R_m\left(\frac{ka^2}{L}, 1\right) \tag{176}$$

Here S_m and R_m are associated wave functions of the first kind and zero order in prolate spheroidal coordinates. They are also called spheroidal wave functions [21]. The field distribution of the lowest order spheroidal wave function is plotted in Fig. 21. The parameter in this diagram is the Fresnel number

$$N = \frac{a^2}{L\lambda} \tag{177}$$

Any optical resonator is characterized completely when the ratio R/L and the Fresnel number N are specified. In the confocal resonator for example with $R = L$ all mode characteristics are just a function of N.

Diffraction losses of the confocal resonator may be determined from its eigenvalues (176). One single transit will change the amplitude by the factor $|\varkappa|$. Therefore the attenuation constant for small losses is

$$\alpha = \frac{1}{L} \ln |\varkappa| \simeq \left(1 - \frac{1}{|\varkappa|}\right) \frac{1}{L}$$

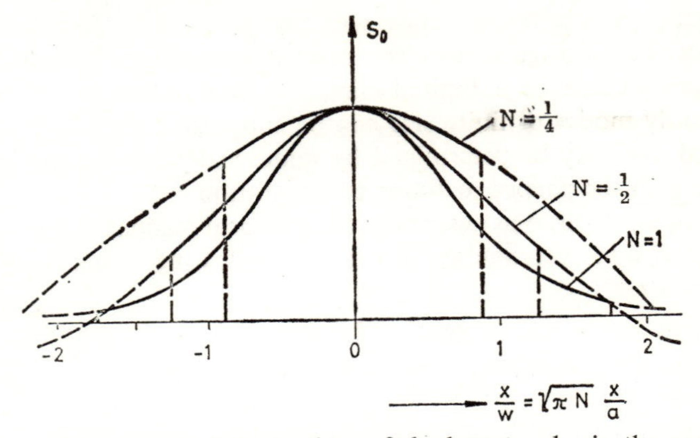

FIG. 21. Transverse field distributions of the lowest order in the confocal resonator for different N. The continuation of the functions beyond the mirror edges have been indicated by broken lines

With $|\varkappa| = |\varkappa_x \varkappa_y|$ we have for the mode of order m in the case of square mirrors

$$\alpha L \simeq 1 - 4NR_m^2(2\pi N, 1) \tag{178}$$

When $N > 0{\cdot}5$ and for the lowest order mode ($m = 0$) this expression is approximated by [22]

$$\alpha L \simeq 4\pi\sqrt{2N}\, e^{-4\pi N} \tag{179}$$

The corresponding approximate formula for the diffraction losses of the lowest order mode is also available in the case of confocal resonators with round mirrors [23]. Here the quantity a in the Fresnel number designates the mirror radius. For large Fresnel numbers the round mirror approximation is

$$\alpha L \simeq 8\pi^2 N e^{-4\pi N} \tag{180}$$

From the optical resonator with infinitely large mirrors we know that for a given resonator length the confocal arrangement amongst all the possible mirror configurations has the smallest beam diameter or spot size. It is obvious then that in the case of mirrors with a limited aperture the confocal arrangement has the smallest diffraction loss. This conclusion has been verified by perturbation calculations [24].

To analyse the modes of non-confocal resonators with mirrors of limited aperture, the corresponding integral equation must be solved numerically. Following the physical process of waves travelling back and forth an iterative procedure is well suited [25]. A field distribution $\mathbf{E}_i^{(1)}$ which comes as close as possible to the actual distribution of a particular mode is assumed on one of the mirrors. For $\mathbf{E}_i^{(1)}$ as a source distribution the field $\mathbf{E}_i^{(2)}$ on the other mirror is calculated by numerically evaluating the integral in (157). This new field distribution will then be resubstituted for $\mathbf{E}_i^{(1)}$ in (153) and the calculation repeated. After a sufficient number of repetitions the distribution will not change much between any two calculations except for a constant factor. Any distributions which these calculations tend towards are modes of the resonator. The constant factor between calculations is the eigenvalue. Mode losses due to diffraction are given by the magnitude of each eigenvalue.

Another procedure for solving the integral equation of the non-confocal resonator with limited aperture starts from a series representation of the unknown modes in terms of the field distribution of modes of the corresponding confocal resonator with limited aperture [24]. The representation, therefore, is in terms of spheroidal wave functions (175). These spheroidal wave functions form a complete and orthogonal set.

The series expansion for the unknown field distributions is introduced into the integral equation (159). By multiplying with spheroidal wave functions of the orthogonal set and integrating over the mirror aperture the integral equation (159) is reduced to a homogeneous system of linear equations for the unknown coefficients in the series expansion. Values of $\varkappa$ for which the coefficient determinant of the homogeneous system will vanish are eigenvalues of the resonator. Once the eigenvalues have been determined the series expansions for field distributions may also be evaluated.

Either the iteration procedure or the normal mode expansion will also serve to analyse the effects of a number of other deviations in the optical resonator from its ideal confocal form. Such deviations include offset mirrors and tilted mirrors as well as spherical aberrations and local variations of the refractive index within the resonator. As a typical example of such calculations we will present only the diffraction losses for the non-confocal but otherwise perfect resonator, and compare them with the characteristics of the confocal resonator. Figure 22 shows such diffraction losses plotted against the deviation $(1-L/R)$ from the confocal system for low order modes in resonators with square mirrors with $N = 1$. The lowest order mode has also the lowest diffraction losses while mode losses increase with mode order. In Fig. 23 the diffraction loss of the lowest order mode is plotted versus $(1-L/R)$ for

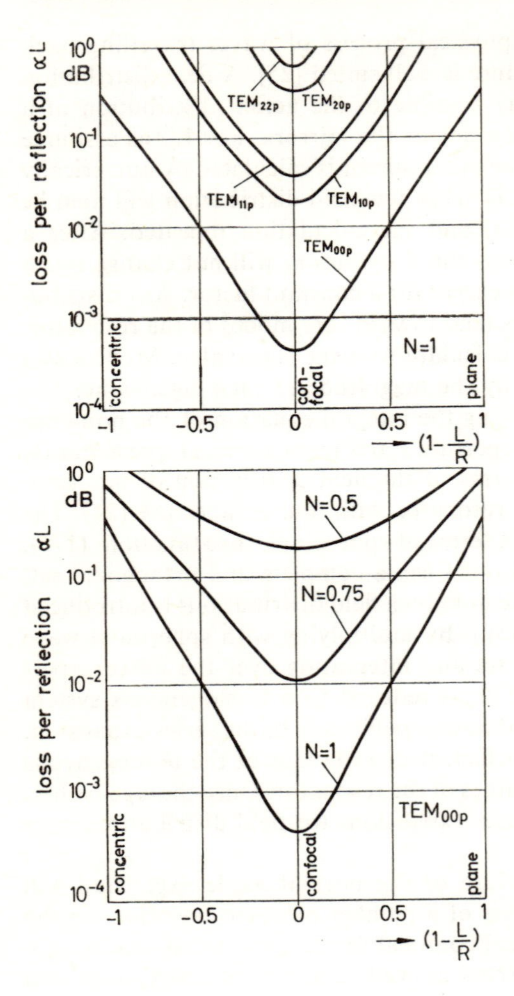

FIG. 22. Diffraction loss of low order modes in the optical resonator with square mirrors and $N = 1$

FIG. 23. Diffraction loss of the lowest order mode in the optical resonator with square mirrors

different Fresnel numbers. With increasing N the minimum loss of the confocal resonator will decrease and be more pronounced.

A typical numerical example for a gas laser serves well to illustrate the characteristics of optical resonators. For a wavelength of $\lambda = 1\ \mu$m and a confocal resonator with $L = 1$ m the spot size on each mirror is

$$w = \frac{1}{\sqrt{\pi}}\,\text{mm}$$

Even if these mirrors were only 2 mm in diameter the diffraction losses according to Fig. 23 would still be only 10^{-4} per reflection. For plane mirrors on the other hand the diameter would have to be at least 6 mm to keep the diffraction loss per reflection below 10^{-2}. In addition these plane mirrors would have to be aligned exactly parallel.

Optical resonators are usually built with interference mirrors. Such mirrors consist of many layers with alternating refractive indices. Each layer is a quarter wavelength thick. The steps in refractive index at boundaries between layers cause partial reflections. All these reflections add in phase while the transmitted wave becomes quite weak. Interference mirrors have selective frequency-dependent reflection. For this reason they are especially well suited for lasers which have active material with stimulated emission in more than one spectral range. In such lasers oscillations will only be generated at the frequencies where the mirror reflection is high.

Interference mirrors are essentially loss-less. All the power of the incident wave is either reflected or transmitted except for losses due to scattering at imperfections in the mirror layers or residual absorption. By a suitable choice of the number of layers they may be designed to have specified reflection or transmission. Even if the only requirement is high reflection, regardless of the transmission, interference mirrors are more efficient than metallic surfaces. The finite conductivity of metals causes reflection losses at optical frequencies to be higher than in multilayer interference mirrors. The following table summarizes the characteristics of high quality interference mirrors [26, 27]:

TABLE 1.

Power Reflection and Transmission of High Quality Interference Mirrors

Layer number	Reflection %	Transmission %	Loss %
7	98·9	1·08	0·02
9	99·4	0·57	0·03
13	99·9	0·06	0·04
30	99·99	0	0·01

Other conditions being equal, mirror loss increases nearly proportionally to the number of layers, except for very many layers. These losses are caused by scattering at layer imperfections and by absorption.

2.9. Beam wave guides

By analogy to the relation between wave guide resonators and the corresponding wave guides there is a wave guide which is related to the optical resonator. This wave guide has favourable dimensions and low loss characteristics at optical frequencies. Its natural modes of propagation may be directly excited by laser beams and it will guide laser beams quite efficiently. For all these reasons this wave guide is of particular interest in quantum electronics.

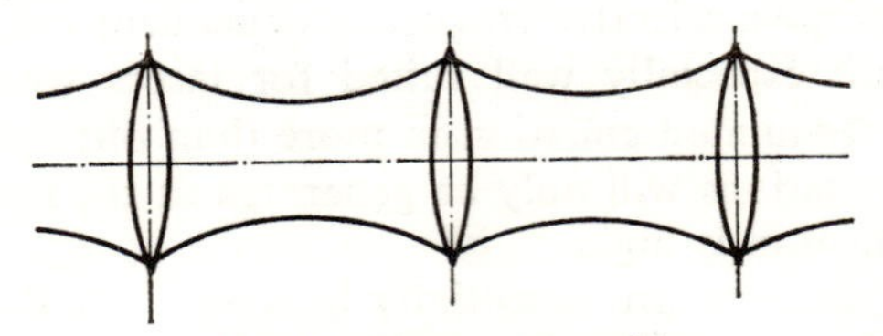

FIG. 24. Beam wave guide

The modes of optical resonators have been interpreted as standing waves which are formed by superimposing two waves travelling in opposite directions. They are waves which, while being continuously reflected by the end mirrors, travel back and forth within the resonator.

If, as shown in Fig. 24, the resonator with spherical mirrors is replaced by a periodic arrangement of lenses of dimensions and focal lengths equal to that of the mirrors the wave will travel through this line of lenses and have the same transverse field distribution as in the resonator. This field distribution will be repeated from lens to lens. In between lenses the field distribution will be somewhat constricted and will spread again at the lenses. To each transverse mode of the resonator there is a corresponding propagating wave guide mode. It has the same diffraction loss per lens as the resonator mode has per reflection.

The diffraction loss per lens for the lowest order mode of propagation may therefore be taken directly from Fig. 23. The propagating modes form well focused beams of light. The periodic guiding structure is therefore referred to as a beam wave guide.

Of all optical resonators the confocal arrangement has the smallest spot size and minimum diffraction loss. The confocal beam wave guide will therefore also be of lowest loss for its propagating modes. In addition to diffraction loss $a_D = \alpha_D L$ any practical beam wave guide has losses $a_\varrho = \alpha_\varrho L$ due to reflection, absorption and scattering at the lenses. The total loss in a beam wave guide with $n = L_T/L$ sections therefore will be

$$a_T = (a_D + a_\varrho)^n$$

For confocal beam wave guides a_D is a function of the Fresnel number $N = a^2/\lambda L$ only. The additional lens losses a_ϱ may be assumed to be a constant quantity which is given by the quality of the material. With $N_T = a^2/\lambda L_T$ the total loss may be written as

$$a_T = \frac{N}{N_T}[a_D(N) + a_\varrho] \tag{181}$$

If for a particular beam wave guide the diameter $2a$ and lens losses a_ϱ are specified then only the spacing of lenses may be varied. Different confocal systems of equal diameter but varying lens spacing will have very many lenses and high lens losses when L is small whereas they will have high diffraction losses when L is large. In between the total loss has a minimum as a function of L or N respectively. This minimum may be obtained from

$$\frac{\partial a_T}{\partial N} = 0$$

Or, using (181), from

$$-N\frac{da_D}{dN} = a_D + a_\varrho \tag{182}$$

In case of small lens losses a_ϱ this equation may be solved approximately for N. For small loss a_ϱ per lens the lenses may be spaced close together. Under these circumstances the minimum total loss will be obtained for large Fresnel numbers. For large N eqn. (180) is in turn a good approximation for the diffraction losses of round lenses. Substituting (180) for a_D in (182) we obtain

$$4\pi N - \ln\left(2\pi - \frac{1}{N}\right) = \ln\frac{16\pi^2}{a_\varrho}N^2$$

Here again because $a_\varrho \ll 1$ we can expect $N > 1$. As long as the results justify this we may therefore calculate N from

$$N = \frac{1}{4\pi}\ln\frac{32\pi^3 N^2}{\alpha_\varrho} \tag{183}$$

This will then be the confocal beam wave guide of minimum total loss in the case of small lens losses.

A numerical example will illustrate how efficient the beam wave guide is for low loss transmission of laser beams. Let the lenses each have 1 per cent power loss due to reflection and other imperfections. With $a_\varrho = 0{\cdot}005$ we obtain from (178)

$$N = 0{\cdot}95$$

It is true that this value for N is somewhat below the limit for (180) to be valid. Nevertheless it will still be a good estimate. If the lenses are 5 cm in diameter and if a laser beam of the widely used helium–neon gas laser is to be transmitted with $\lambda = 0{\cdot}63$ μm then the lens spacing should be $L \simeq 1$ km. The total loss under these optimum conditions is only 0·045 db/km. Lens losses of $\alpha_\varrho = 0{\cdot}043$ db/km constitute the major part.

As a whole the loss is one to two orders of magnitude smaller than that for transmission lines or wave guides at lower frequencies. The installation of such beam wave guides would be quite difficult however, since the sections between lenses must be optically straight. Mirror deflectors or prisms must be used for changes in the direction of propagation, and, in addition, the lenses must be carefully aligned and kept in a stable position.

Variations in the medium between the lenses must be excluded by placing the beam wave guide in a protective pipe which may possibly have to be evacuated or filled with an inert gas. Automatic controls are also under consideration which continuously monitor the beam position and keep it centred by compensating with beam deflectors for any changes.

Considerable research effort is being expended on the beam wave guide since it seems to be a particularly attractive medium for signal transmission via laser beams.

2.10. General requirements for laser media

Based on our discussion of all the physical phenomena which bring about and influence laser action it can now be decided which media are suitable for lasers and what requirements they should meet. Of all media which exhibit fluorescence those which fit the following demands will be particularly efficient when used in lasers.

1. The lower state 1 of the induced transitions should be high enough above the ground state for its population at thermal equilibrium to be small compared with the population inversion required for laser action ($N_1 < \Delta N$). In other words: four-level laser operation is desirable.
2. The line width Δf of the induced emission should be small so that high gain may be achieved or the threshold value for the fluorescence power may be kept low.
3. The upper state 2 of the induced emission should be metastable. Its lifetime τ_2 should be long and at best be limited only by spontaneous transitions to the lower state 1 of the induced emission ($\tau_2 = \tau_{12}$).
4. The lifetime τ_1 of the lower state 1 should be short. τ_1 should, in any case, be smaller than τ_2.
5. The laser material should have no loss in the range of the transition frequency f_{12} other than that due to induced absorption of the transition $1 \rightarrow 2$. All other losses would increase the attenuation α and require the population inversion ΔN to be correspondingly higher.
6. In general the laser material should be of high optical quality and have chemical as well as thermal and mechanical stability.
7. For optical pumping at frequency f_{03} the medium should have efficient absorption in a wide spectral range near f_{03}.
8. The fluorescence efficiency should be high. For nearly every photon which is absorbed at the pump frequency one fluorescence photon should be emitted.

2.11. Solid-state laser material

Most of the requirement for efficient laser action are met quite well by crystals or glasses which have been doped with small amounts of active elements. These doping elements should have only partially populated inner electron shells with transitions between them. The outer electrons will then screen these inner shells and reduce the interaction of inner electrons with the surrounding crystal. As a consequence the emission lines are very narrow. Doping elements which have these characteristics are metals of the transition group, rare earth elements, and elements of the actinide series.

If *crystals* are doped with such elements the emission lines will be broadened mainly thermally through interaction with lattice vibrations. The thermal energy of active micro-systems will change with time and the transition frequency may shift throughout the entire width of the emission line. The time intervals between such energy fluctuations are determined by acoustic velocities. They are only of the order of 10^{-13} sec. The photon lifetime is normally much longer for a particular mode of the resonator. During this lifetime, and due to the rapid thermalilization, nearly all micro-systems will be tuned temporarily to the transition frequency of one particular photon within the emission linewidth. Since the energy levels are scattered over the linewidth each photon will have equal likelihood of inducing transitions in a micro-system. Relative to the long-living photons the emission line in crystals is homogeneously broadened and when oscillations are excited it will be homogeneously saturated. In multimode operation the envelope of the energy distribution among the modes will be narrowed. With increasing pumping power the energy will be concentrated more and more in modes near the line centre.

If *glasses* are doped with active elements the emission line will be broadened by the random nature of the surrounding glass. In this case the deviation of energy levels from an average level will not change with time. Photons in one particular mode will be likely to induce transitions in only a fraction of the micro-systems. Thus in this case the emission line is inhomogeneous and will be selectively saturated.

Population inversion is generated in solid state laser materials by absorption of electromagnetic radiation in a relatively wide band of frequencies and subsequent non-radiative transitions to the narrow levels of inner shells of the active doping element.

2.11.1. Transition Metals in Crystals (Ruby)

Laser action was demonstrated for the first time in single crystals of Al_2O_3 which were doped with Cr^{3+} ions. This system is commonly known as ruby. The energy level diagram is shown in Fig. 25. The various levels have been designated in accordance with their strong dependence on the crystalline electric field.

Emission is generated by induced transitions from the ^2E-states to the $^4A_2$ ground state. We therefore have a three-level laser. Upon closer examination we find that the ground state is split into two levels. Their separation is small enough however for the upper of both levels to be heavily populated even at very low temperatures.

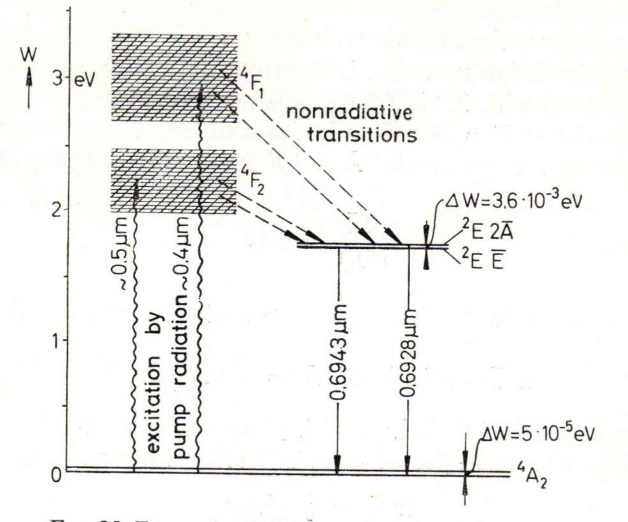

FIG. 25. Energy-level diagram of Al_2O_3: Cr^{3+} (ruby)

Normally emission is observed at $\lambda = 0{\cdot}6934$ μm corresponding to the transition $\bar{E} \rightarrow {}^4A_2$. Emission due to the transition $2\bar{A} \rightarrow {}^4A_2$ at $\lambda = 0{\cdot}6919$ μm is always weaker. Because of the rapid thermalization $2\bar{A}$ is less populated than $\bar{E}$ and the metastable lifetime in $\bar{E}$ is larger. Oscillations at the transition frequency of $2\bar{A} \rightarrow {}^4A_2$ will therefore be suppressed by oscillations due to $\bar{E} \rightarrow {}^4A_2$.

By observing the decay time of fluorescence a lifetime of $\tau_2 = 3$ msec was found for the metastable state $\bar{E}$. This lifetime is mainly limited by spontaneous transitions to ${}^4A_2$ and not so much by nonradiative transitions of thermalization. We therefore have $\tau_{12} \simeq \tau_2 = 3$ msec.

Population inversion is generated by absorption of pumping power and excitation into the energy bands ${}^4F_2$ and ${}^4F_1$ with subsequent non-radiative transitions to ${}^2E$-states. The efficiency of fluorescence is observed to be nearly unity. There are far less spontaneous or non-radiative transitions, e.g. from ${}^4F_2$ to ${}^4A_2$, than non-radiative transitions from ${}^4F_2$ to ${}^2E$.

The line width of induced transitions $\bar{E} \rightarrow {}^4A_2$ at room temperature is $\Delta f = 33 \times 10^{11}$ Hz corresponding to $\Delta\lambda/\lambda = 0{\cdot}076$ per cent. At 77°K it even narrows to $\Delta f = 3 \times 10^{10}$ Hz corresponding to $\Delta\lambda/\lambda = 7 \times 10^{-6}$. Ruby crystals for laser action are normally selected for high optical quality from carefully grown lots.

In many of its characteristics ruby meets most of the requirements for efficient laser action. However, it is not a four-level laser. The pump power for population inversion is therefore quite high. A typical ruby crystal is doped with 2×10^{19} Cr atoms per cm^3. According to Boltzmann statistics at $T = 290°K$ and with a difference $\Delta W = 0{\cdot}0036$ eV between the two upper levels $2\bar{A}$ and $\bar{E}$ the density of states are in the ratio

$$\frac{n(2\bar{A})}{n(\bar{E})} = 0{\cdot}87$$

The sum of all state densities must be equal to the doping density of Cr atoms

$$n(\bar{E}) + n(^4A_2) + n(2\bar{A}) = 2 \times 10^{19} \text{ cm}^{-3}$$

The states in energy bands $^4F_2$ and $^4F_1$ have short enough lifetimes for their densities to be neglected here. The critical inversion requires

$$n(\bar{E}) = \tfrac{1}{2} n(^4A_2) \tag{184}$$

The factor $\frac{1}{2}$ accounts for the degeneracy in state $^4A_2$ which is twice the degeneracy of state $\bar{E}$. For the critical inversion we obtain from these three relations

$$n(\bar{E}) = 5{\cdot}2 \times 10^{18} \text{ cm}^{-18} \tag{185}$$

and hence a threshold power for fluorescence from (122)

$$\frac{P_f}{V} = 615 \frac{\text{W}}{\text{cm}^3} \tag{186}$$

Here we have assumed all excited micro-systems to be populated from the band $^4F_2$ and have used $\lambda_{03} = 0{\cdot}56$ μm as the pumping wavelength. Furthermore, we have not considered any of the losses which must be overcome by increased inversion of states. Even so the threshold power of (186) is already quite high. A substantial part of this pumping power will generate heat during the non-radiative transitions from $^4F_2$ to ^2E. Some of the pumping power will not even be useful in exciting states in $^4F_2$ and $^4F_1$ and it is also dissipated in heat. In all so much heat is generated that cw operation requires efficient and careful cooling. With such cooling for efficient heat dissipation and well focused pumping sources cw operation is possible at 77°K as well as at room temperature [31].

On the other hand, the ruby laser is very well suited for pulsed operation.

2.11.2. Rare Earths in Crystals

In rare earth ions for lasers the electrons of the 4*f*-shell make induced transitions. The interaction between the host crystal lattice and these levels is weak enough for them to be designated by the corresponding levels of the free ion. This designation is different from the level designation for transition metals. Stimulated emission is due to transitions between states with $n = 4$ and $l = 3$ of the 4*f*-shell. Electrons of these

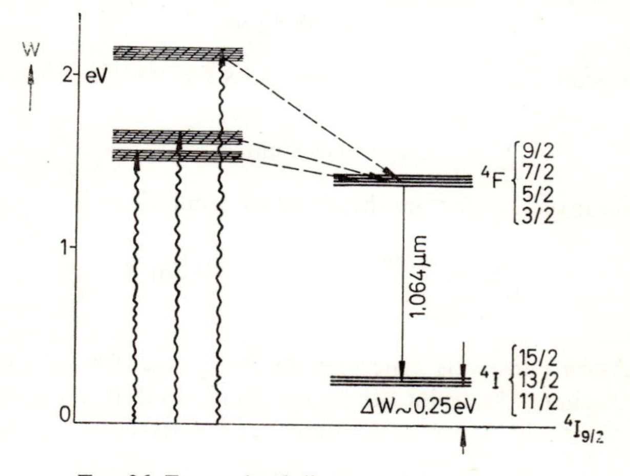

Fig. 26. Energy-level diagram of $CaWo_4$: Nd^{3+}

states are well shielded from the lattice field by two 5*s*- and six 5*p*-electrons. Their levels are nearly independent of the particular host crystals. Transitions between them are distinguished by narrow lines and are predominantly of a spontaneous or induced nature. The divalent rare earth ions have strong absorption bands in the visible range of frequencies due to $4f-5d$ transitions, which may serve to excite them. The terminal states of the stimulated emission in the 4*f* shell have high enough levels and possibly allow four-level laser operation with cooling. Pumping power is relatively low and cw-operation feasible.

Two materials of this class, Nd^{3+} in $CaWo_4$ and Dy^{2+} in CaF_2 were first used for cw-lasers. We will only discuss Nd^{3+} in $CaWo_4$. Its energy-level diagram is shown in Fig. 26. Infrared fluorescence is observed due to transitions from $^4F_{\frac{3}{2}}$ to the various levels of the ground ^4I multiplet. Fluorescence at $\lambda = 1{\cdot}065$ μm from the $^4F_{\frac{3}{2}} \rightarrow {}^4I_{\frac{11}{2}}$ transition is strongest and will lead to laser action first. The terminal level $^4I_{\frac{11}{2}}$ is $\Delta W = 0{\cdot}25$ eV above the ground state. Four-level laser operation is therefore feasible even at room temperature. The emission line is nearly

$$\Delta f = 2\times 10^{11}\ \text{Hz}$$

wide.

From relaxation measurements the lifetime of the upper laser level was found to be

$$\tau_{12} = 10^{-4}\ \text{sec}$$

If for a laser of 3 cm length the power loss per transit is 5 per cent we have

$$\alpha = 0{\cdot}008\ \text{cm}^{-1}$$

The critical inversion for the threshold of oscillations is

$$\Delta n = \frac{\alpha \Delta\omega}{\pi c^2}\,\omega_{12}^2\tau_{12} \simeq 10^{-15}\ \text{cm}^{-3}$$

This inversion density is generated by pumping with a high-pressure mercury discharge lamp into the absorption band at $\lambda_{03} = 0{\cdot}57 - 0{\cdot}60$ μm. The threshold power of fluorescence is determined from (123) as

$$\frac{P_f}{V} = 4\,\frac{\text{W}}{\text{cm}^2}$$

It is considerably lower than that for the three-level ruby laser. CW-operation is feasible without too much difficulty. Actually 1 W of cw output power has been obtained with water cooling only. Even higher cw power at room temperature has been generated with Nd^{3+} in $Y_3Al_5O_3$ (yttrium aluminium garnet).

2.11.3. Ions of the Actinide Series in Crystals

The atoms of the actinide series have unpaired electrons of the 5*f*-shell which are only partially shielded by outer 6*s*- and 6*p*-electrons from the lattice field of a host crystal. The shielding is not as effective as it is for the 4*f* electrons in rare earth atoms but nevertheless the emission lines for transitions of the 5*f* shell in actinide ions are still quite narrow.

Uranium and thorium are the only two elements of this series which do not show radioactive decay. Of these two U^{3+} in CaF_2 was first observed to exhibit laser action. The relevant energy levels are displayed in Fig. 27.

Transitions which start from a state of the $^4I_{\frac{11}{2}}$ multiplet and terminate in a state of the $^4I_{\frac{9}{2}}$ multiplet may be stimulated to emit

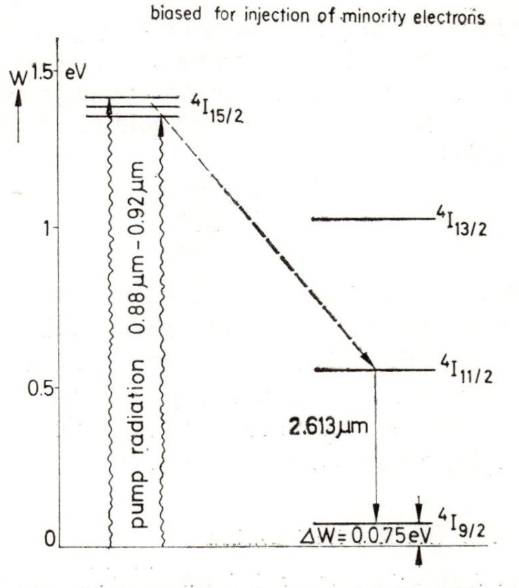

FIG. 27. Energy-level diagram of CaF_2: U^{3+}

radiation at $\lambda = 2{\cdot}61\ \mu$m. The terminal level is $\Delta W = 0{\cdot}075$ eV above ground level. Cryogenic cooling to below 100°K will ensure four-level laser operation.

At low temperatures the initial state of the $^4I_{\frac{11}{2}}$ multiplet is metastable with a lifetime of 130 μsec. At higher temperatures the lifetime is shortened by non-radiative transitions to 15 μsec at 300°K. The transition for stimulated emission has a line width of

$$\Delta f = 6 \times 10^{11} \text{ Hz}$$

at 77°K.

For a laser of 3 cm length and a power loss of 5 per cent per transit we have

$$\alpha = 0{\cdot}008\ \text{cm}^{-1}$$

To overcome this loss and excite oscillations requires an inversion density from (67), (116) and (121) of

$$\Delta n \simeq 7\times10^{14}\ \text{cm}^3$$

This critical inversion may be obtained best by pumping in the wavelength range $\lambda = 0{\cdot}88$–$0{\cdot}92\ \mu$m. There are three absorption bands in this range. A xenon lamp is suitable for this optical pumping. Assuming perfect efficiency of fluorescence the threshold for the power of fluorescence from (123) is

$$\frac{P_f}{V} = 1{\cdot}2\,\frac{\text{W}}{\text{cm}^3}$$

This threshold power is still lower than is the case for Nd^{3+} in $CaWO_4$. On the other hand, however, Nd^{3+} absorbs pumping radiation more efficiently than U^{3+}. For this reason nearly the same pumping power is required in both cases. In a typical experiment at 77°K a pumping power of 250 W was required to start cw-oscillation while for 700 W pumping power the cw output power was 1 W.

2.11.4. Glass and Glass Fibre Lasers

A number of rare earth elements with Nd, Yb, Ho and Er among them can be made to lase in glass as the host material. Instead of the unique and well-defined crystal lattices the active atoms are now incorporated into a matrix which is inherently inhomogeneous and differs between micro-systems. The glass host produces inhomogeneously broadened lines which are wider than would be found in crystals. The threshold for fluorescence therefore is harder to obtain because a larger inversion is required than for the same active ion in a host crystal.

Glass may be obtained with excellent optical quality however, and affords considerable flexibility in size and shape. Oscillations may be excited by laser action in very thin active glass fibres of a few microns diameter which are surrounded by a passive glass coating of lower refractive index. The external cladding dissipates heat and the thin fibres support well defined surface waves as modes of propagation and oscillation.

Glass lasers can also be made as rods a few centimetres in diameter and up to 2 metres long. Because of this large size and their broadened fluorescent line such glass lasers are well suited for high-energy pulsed operation.

Of the five rare earth elements which have been listed above neodymium is most widely used in glass lasers. It has also been made to lase in a large variety of glasses. Usually the emission from Nd^{3+} is at 1·06 μm. Not only pulsed but also cw-operation is possible.

2.12. Semiconductor lasers

If minority carriers are injected into highly doped semiconductors they may undergo a recombination transition and thereby radiate stimulated emission. In the case of minority electrons the recombination transition starts from the lower edge of the conduction band or from donor levels just below the edge of the conduction band and terminates at the upper edge of the valence band or in shallow acceptor levels. In all cases the energy difference of the band gap is emitted as a photon and determines the radiation frequency.

A *p–n* junction biased into conduction normally serves to inject minority carriers. One side of the *p–n* junction must be so highly doped that the Fermi level is very close to or even within the band. The semiconductor is then nearly degenerate or even reaches degeneracy. In the case of a degenerate *p*-region as in Fig. 28 the acceptor levels and shallow levels in the valence band are empty.

Minority electrons which are injected from the *n*-region through the *p–n* junction at first will stay in the conduction band and populate low-lying levels in this band as well as donor levels. The population inversion is thus as occurs in the four-level laser.

Not all transitions between states in the conduction and valence band are allowed, however. For such a transition to be likely it must be between states of equal momentum. Electrons at the lower edge of the conduction band must have the same momentum as holes at the upper edge of the valence band. Only under these circumstances will momentum be preserved when the electron makes a direct transition for recombination. Otherwise the transition cannot be direct but must be indirect by requiring a lattice vibration phonon to absorb the difference in momentum. Such indirect transitions where phonons are involved are so unlikely that they may be classified as not allowed.

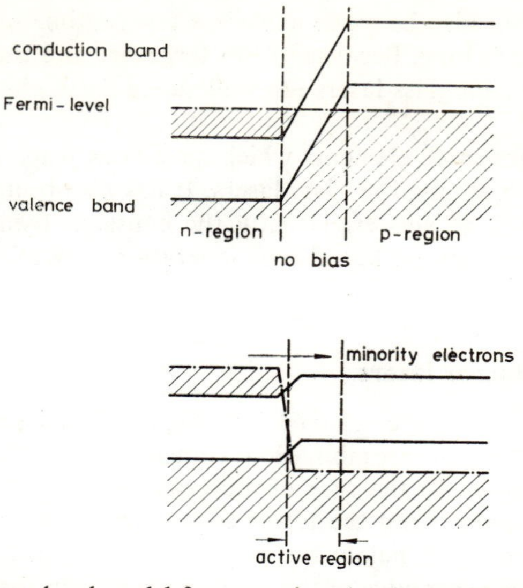

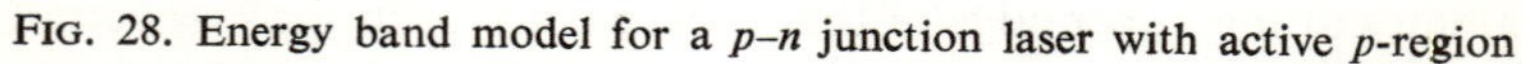

FIG. 28. Energy band model for a *p–n* junction laser with active *p*-region

Direct transitions exist in such III–V compounds as GaAs, GaP and GaSb. In silicon, on the other hand, transitions from band to band are indirect. They are very unlikely and this material is not suitable for semiconductor lasers.

The transitions between the well-defined levels of donors and acceptors result in relatively narrow emission lines. Between other electronic states we have a continuous spectrum which is unique for semiconductor lasers. Loss of radiation is mainly caused by free carriers absorbing energy. In the active region adjacent to the *p–n* junction this loss is overcome by induced emission. Modes of propagation are excited within the active region which corresponds to the surface waves in dielectric slabs. They travel parallel to the *p–n* junction. For these waves to form oscillatory modes the two opposing faces must be cut parallel and polished. Mirror coatings of these faces or reflectors will enhance oscillations.

Significant advantages of semiconductor lasers are:

1. They are directly excited by a d.c. current driven through the *p–n* junction.
2. They posses high efficiency: with quantum efficiencies of nearly 100 per cent the energy of each injected minority carrier appears

as radiation energy. Nearly all of the d.c. power which is supplied by a bias source will therefore be converted into coherent radiation power.

3. Rapid switching and amplitude modulation is easily effected by simply controlling the bias voltage at the *p–n* junction.
4. Because of small dimensions and low power requirements the semiconductor laser is a very flexible and handy device.

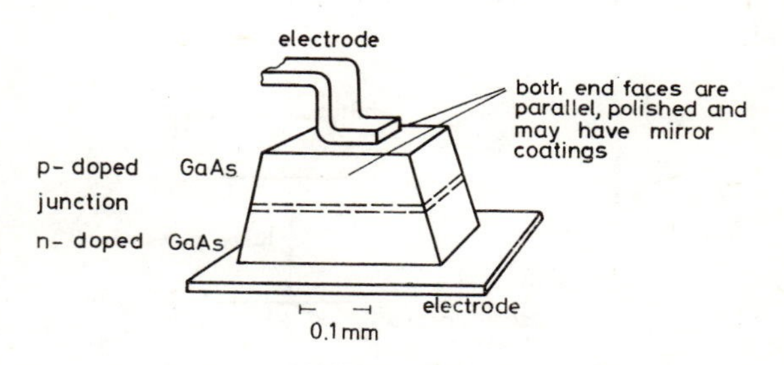

FIG. 29. GaAs diode as injection laser

Intense recombination radiation from the injection of minority carriers is observed in GaAs, GaP and GaSb. In particular GaAs-diodes (Fig. 29) show efficient laser action. A degenerate *p–n* junction in GaAs may be fabricated by diffusing Zn into Te-doped material. Recombination of minority electrons and radiation occurs in the predominantly Zn-doped *p*-region of the junction. The threshold of stimulated emission requires a current density of 10^9 A/cm^2 at 77°K. The radiation at $\lambda = 0{\cdot}84\ \mu$m corresponds to the energy gap of GaAs at this temperature.

Because of the high current density for stimulated emission only pulsed operation was feasible at first. The current density at threshold is less for even lower temperature. At 4·2°K a current density of only 700 A/cm^2 is sufficient. At this temperature GaAs lasers are operated cw and generate a few watts of output power. At room temperature with pulsed operation a peak output power of 10 W has been obtained.

As the technology is improved semi-conductor lasers are expected to become more efficient and work continuously even without cryogenic cooling. For pulsed operation the peak output power is expected to be raised to 10 kW. It is quite likely that for many future practical applications the semiconductor laser will be of the greatest significance among all lasers.

2.13. Gases and gas mixtures

There are also metastable states of atoms and molecules in gases. These states are depopulated only by spontaneous or induced emission and if their density is sufficiently high this may lead to laser action. To achieve such a high population they are excited either directly or indirectly through collisions with free electrons in the gas discharge. The gas discharge in turn is maintained by a radio-frequency or d.c. electric field.

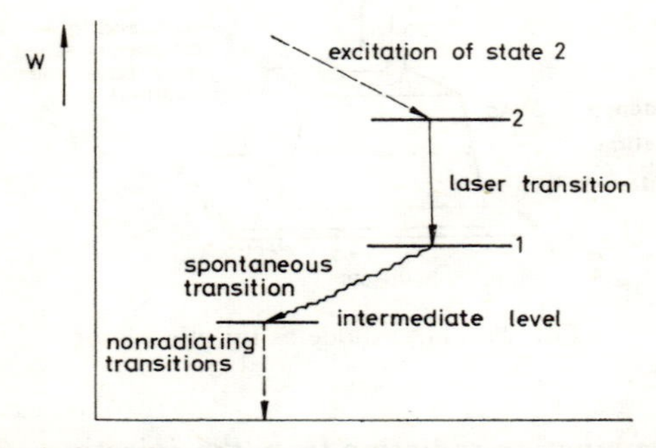

FIG. 30. Depopulation of the lower laser level by spontaneous transitions to an intermediate level

Most of the gas lasers work with a four-level energy scheme. To generate and maintain the required inversion density in such a four-level gas laser the terminal state ofinduced transition must be depopulated efficiently. While in solid state lasers depopulation of the lower level by interaction with the surrounding lattice or matrix is usually quite effective, it requires special attention in gas lasers. In gas lasers such lower levels are normally also metastable. Even if their rate of spontaneous transitions to the ground state is very high the spontaneous photons are very likely to be trapped by one of the many micro-systems in the ground state. Such a micro-system is then excited into the terminal state and thus curtails the depopulation of this state. There are two ways to get round this obstacle to high inversion density:

1. If micro-systems with energy levels as shown in Fig. 30 are used the intermediate level will help to depopulate the terminal level of the laser transition. The density of states in the intermediate level is ordinarly very low according to Boltzmann statistics. Many micro-systems will therefore make spontaneous transitions to this state with-

out any radiation trapping. An additional requirement for these systems is that subsequently they leave the intermediate level by non-radiative transitions to the ground state. A high rate for the depopulation of the intermediate level may be achieved by intimate contact of the gas with solid walls and the ensuing relaxation due to wall collisions. For this purpose the discharge tube is made of small diameter.

2. The other way to depopulate the lower laser level is to keep the ground level empty. Low density of ground states will be obtained in

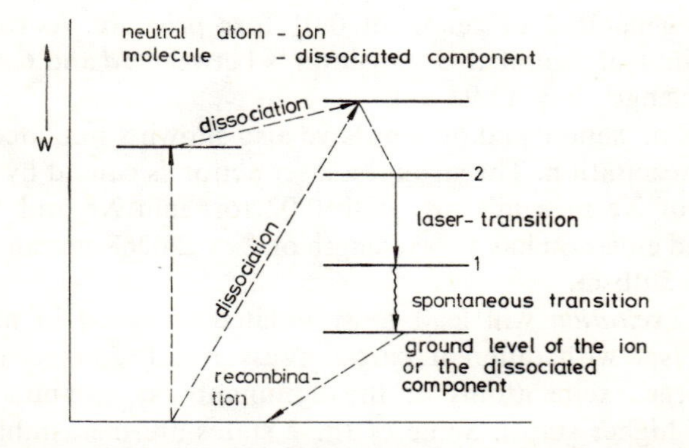

FIG. 31. Depopulation of the lower laser level by spontaneous transitions to a weakly populated ground state

ionized atoms or in monatomic components of a dissociated molecular gas. The energy level diagram for such a system is shown in Fig. 31. The laser transition is between states of the ion or the dissociated component. Its terminal state discharges spontaneously to the ground state. From this ground state, the ions or dissociated components recombine at a relatively fast rate to form neutral atoms or molecules.

The excitation of micro-systems into the initial and upper state of the laser transition in gases may also be effected by two different mechanisms. One is a *direct excitation,* the other an indirect excitation.

Direct excitation is achieved by inelastic collisions with free electrons of the gas discharge. The collision cross-section for excitation of a given state is proportional to the dipole matrix element connecting the excited state with the ground state. For any two states the corresponding state functions define the matrix element according to (42). The collision cross-section depends on the magnitude of the matrix element, so that certain states may be excited selectively.

Laser action through this mechanism of direct excitation has been observed in discharges of pure noble gases such as He, Ne, Ar, Kr or Xe. The emission lines are all in the infrared.

The lower laser level for such directly excited gas lasers is always depopulated according to Fig. 30 by spontaneous transitions to an intermediate level and subsequent non-radiative transitions to the ground state. The diameter of the discharge tube must be small to ensure efficient wall relaxation.

Oscillations of the longest frequency in such pure noble gas lasers have been generated in xenon of 0·01 torr pressure. As for all other emission lines of xenon this transition is between $5d$ and $6p$ levels and has a wavelength $\lambda = 12{\cdot}92\ \mu$m.

Mixtures of xenon and helium have also shown stimulated emission with direct excitation. The strongest laser action is caused by a $5d \rightarrow 6p$ transition of Xe in a mixture with 0·02 torr mm Xe and 5 torr He. The induced emission has a wavelength of $\lambda = 2{\cdot}0268\ \mu$m and produces a gain of 4·5 db/m.

Indirect excitation will lead to stimulated emission in mixtures of different gases with different partial pressures. Free electrons of the gas discharge excite atoms of the dominant gas component into a number of higher states. Some of these states are metastable and will become more densely populated. In a suitable combination of gases such metastable states have nearly the same energy levels as metastable states of the weaker gas component. By colliding with atoms or molecules of the weaker component these states will therefore be excited in the weaker component while the collision partners of the stronger component drop back to ground level. For a small difference in levels such an energy exchange by collision is very likely. The small energy difference will be converted into thermal motion.

The weaker component of the gas mixture is thus indirectly and selectively excited into specific metastable states. For transitions from these states to lead to laser action lower lying levels must have a sufficiently short lifetime and must be accessible by induced transitions.

The He–Ne laser is a typical example. The energy-level diagram of a He–Ne mixture is shown in Fig. 32. Energetic free electrons excite helium atoms into a number of higher levels from which they cascade down again. Many of them will remain in the metastable states 2^1S and 2^3S for some time. Here they are likely to collide with neon atoms and raise them into states of the $3s$ and $2s$ multiplets which have nearly the same energy. From these states which in turn are also metastable the neon laser transition will be induced. The terminal states are of the $3p$ or $2p$ multiplets. They discharge by spontaneous transitions to the $1s$ states. Thermal interactions with the walls of the discharge tube will subsequently let the neon atoms return to the ground state. Participating states in the $3s$ and $2s$ multiplets have a lifetime of 0·1 μsec while in the $2p$ multiplet it is only 0·01 μsec. The population is therefore easily inverted by this indirect excitation. However, the $1s$ levels are also metastable. During laser action their population can grow dense enough

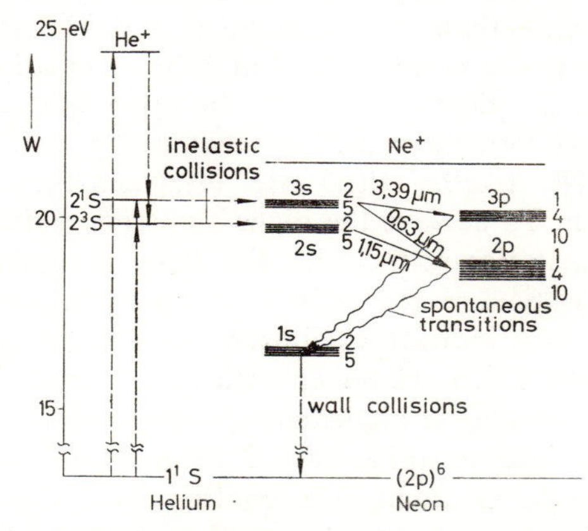

FIG. 32. Energy-level diagram of helium and neon

to trap the spontaneous radiation of $2p \rightarrow 1s$ transitions and by induced $1s \rightarrow 2p$ transitions obstruct the depopulation of the $2p$ states, thus effectively raising their lifetime. In a pure He–Ne mixture the neon atoms will only leave the $1s$ states by thermal interaction with the walls. The mechanism for depopulation of the lower laser level corresponds to that in Fig. 30 with an additional intermediate level. The cross-section of the discharge tube must therefore be small for effective diffusion of neon atoms to the walls.

Neon atoms of the He–Ne mixture will also be directly excited through collisions with free electrons. This direct excitation will not select the initial state of the transition as against the terminal state. It will therefore impair population inversion. For the direct excitation of neon atoms to be weak compared with selectively indirect excitation the He component of the gas must dominate the Ne. Optimum laser action is obtained for the following ratio of partial pressures:

$$p_{\text{He}} = 5p_{\text{Ne}}$$

Stimulated emission in He–Ne is quite strong at $\lambda = 1{\cdot}15259$ μm due to the $2s_2 \rightarrow 2p_4$ transition. In the visible range of the spectrum the most powerful oscillation is at $\lambda = 0{\cdot}6328$ μm due to the $3s_2 \rightarrow 2p_4$ transition. CW oscillation of several 100 mW of output power have been obtained at this wavelength.

An additional laser line is at $\lambda = 3{\cdot}39$ μm due to the $3s_2 \rightarrow 3p_4$ transition. The small signal gain of this infrared wavelength is very high at $2v = 20$ db/m. Under similar circumstances the gain at $\lambda = 0{\cdot}63$ μm is only $2v = 0{\cdot}2$ db/m. The induced emission at $\lambda = 3{\cdot}39$ μm provides so much more gain than the emission at $\lambda = 0{\cdot}63$ μm because the density of states in the terminal level of $3p$ is so much less than in the terminal level of $2p$. Equal density of the upper states in $3s$ means a much higher inversion at $\lambda = 3{\cdot}39$ μm than at $\lambda = 0{\cdot}63$ μm. Both lines have the same Ne state as initial laser level and they interact during induced emission. To achieve the highest possible gain for just one of these lines the excitation of the other line must be suppressed. Selective mirrors for the optical resonator will serve this purpose.

Laser action with indirect excitation has also been observed with oxygen molecules which are mixed with either neon or argon. Neon or argon atoms in states of higher energy collide with oxygen molecules and excite them into unstable states. From these states the oxygen molecules split into O atoms at ground level and excited O atoms in the 3^3p_2 state. A subsequent transiton of excited O atoms from 3^3p_2 to 3^3s leads to stimulated emission.

Another example of indirect excitation is the CO_2 laser. Here carbon dioxide is mixed with helium and nitrogen. The discharge excites N_2 molecules into metastable states. Colliding selectively they invert the population of rotational transitions in CO_2 molecules at $\lambda = 10{\cdot}5$ μm. The excitation is very efficient for reaching high inversion densities. CO_2 lasers can be designed to generate several hundred watts of output power with efficiencies of more than 10 per cent with respect to the d.c. power fed into the gas discharge.

There is still another group of lasers with excitations which must be classified as indirect. They are the ion lasers. The most important example of this group is the argon laser. But ion lasers also work with other noble gases. Laser transitions in this group are between the excited states of ionized atoms of the noble gases. During the process of excitation the atom is first ionized by electron collision, a second collision will then transfer it to an excited state. Sometimes even one collision will produce the excited ion. The lower laser level is depopulated according to Fig. 31 by spontaneous transitions to the ground state of the ion where the population is low. A high rate of recombination for the ions at ground level furthers a low density in the terminal state of the laser transition. High inversion densities in ion lasers lead to large gain factors. An argon laser may have a gain of more than 10 db/m at $\lambda = 0{\cdot}488\ \mu m$.

The preceding survey of mechanisms for excitation and discharge in gas lasers is complete as a whole, but has not mentioned all details. In particular only the most important types of gas lasers have been listed in accordance with their mechanism of excitation. Altogether more than a thousand different emission lines have been observed in gases and gas mixtures. For most of them the energy levels for the induced transition of the particular atoms or molecules have been identified. The highest frequency laser action in gases was observed in the ultraviolet region at $\lambda = 0{\cdot}26249\ \mu m$ for the argon isotope IV. The lowest frequency laser action was measured at $\lambda = 337\ \mu m$ in a mixture of HCN, CH_3CN and C_6H_6CN. This lowest frequency is already in the submillimetre range. It almost closes the gap between lasers and gas masers.

Emission lines of gas lasers are broadened due to the finite lifetime of the states as limited by spontaneous transitions, and due to the Doppler effect of random thermal motion. Finite lifetime broadens the normalized line shape to a Lorentz curve. If spontaneous transitions occur only between the two laser levels 2 and 1 and the lifetime is limited to τ_{12} then the 3-db width of the Lorentz curve is given by

$$\Delta f = \frac{1}{2\pi\tau_{12}}$$

If in addition there are transitions to other states the 3-db width is better expressed in terms of the lifetime τ_1 in state 1 and the lifetime τ_2 in state 2:

$$\Delta f = \frac{1}{2\pi}\left(\frac{1}{\tau_1}+\frac{1}{\tau_2}\right)$$

The time constants τ_1 and τ_2 are not always determined by spontaneous transitions alone. Lifetime in certain states may be limited by collisions between atoms or molecules as well as with the walls of the discharge tube. For most gas lasers under normal conditions the lifetime is in the range between 10^{-8} and 10^{-7} sec. We therefore have a natural line width of 10–100 MHz. The line broadening due to limited lifetimes is homogeneous. For sufficiently strong induced emission the line will be homogeneously saturated.

The line shape due to the Doppler broadening of random thermal motion may be calculated by assuming a Maxwellian distribution for the random velocities of atoms or molecules in the gas. Of the total velocity only the component v parallel to the direction of propagation of the stimulating wave will give a Doppler shift to the interaction. In the case of a Maxwellian distribution we have for the probability dw of a particle travelling in a specific direction with a velocity between v and $v+dv$ [28]

$$dw = \frac{1}{\sqrt{\pi}\, v_T} e^{-\left(\frac{v}{v_T}\right)^2} dv$$

Here v_T is the r.m.s. value of velocity which comes from the average kinetic energy according to

$$\frac{m v_T^2}{2} = kT$$

as

$$v_T = \frac{2kT}{m} \tag{187}$$

where m is the mass of the particle and k the Boltzmann constant. The Doppler effect will shift the transition frequency of a particle at speed v by

$$\frac{f - f_0}{f_0} = \frac{v}{c}$$

Therefore we obtain for the probability dw that a micro-system will have a transition frequency between f and $f+df$ for the interaction with a wave in the direction of v:

$$dw = \frac{c}{\sqrt{\pi}\, v_T f_0} e^{-\left(\frac{f-f_0}{f_0} \frac{c}{v_T}\right)^2} df$$

The coefficient of df in this expression is the normalized line shape

$$g(f) = \frac{c}{\sqrt{\pi}\, v_T f_0} e^{-\left(\frac{f-f_0}{f_0} \frac{c}{v_T}\right)^2}$$

$g(f)$ is a Gaussian curve. Comparing it with (93) the 3-db bandwidth is found to be

$$\Delta f = 2f_0 \sqrt{\frac{2kT}{mc^2} \ln 2} \tag{188}$$

The Doppler width increases linearly with frequency; for higher atomic or molecular weight on the other hand the Doppler width will be smaller. In the He–Ne laser with the atomic weight m of neon in (188) the line at $\lambda = 0{\cdot}63\ \mu m$ is

$$\Delta f = 1500 \text{ MHz}$$

wide. In this case as in many other gas lasers Doppler broadening dominates the natural line width.

The Doppler effect line broadening is inhomogeneous under normal conditions. Normally the lifetime of active micro-systems is much longer than the lifetime τ_p of photons in optical resonator modes. When these modes are excited they will therefore selectively saturate the Doppler broadened line. According to (155) only a fraction of the active micro-systems will be available for induced emission into a particular mode. The threshold power for fluorescence will now be calculated for a representative example. This will serve to estimate the power requirement for gas lasers: the density of states is relatively low in low-pressure gases; gain constants are also small. Sufficient gain for the excitation of oscillations is obtained when discharge tubes are about 1 m long. Proper design of the resonator will keep reflection and diffraction power losses below 2 per cent per transit. To excite oscillations the gain must overcome

$$\alpha = \frac{a}{L} = \frac{0{\cdot}01}{100} = 10^{-4} \text{ cm}^{-1} \tag{189}$$

of attenuation. A suitable He–Ne mixture has a lifetime

$$\tau = \frac{1}{A} = 10^{-7} \text{ sec}$$

for the upper laser level in the $2s$-multiplet. Let oscillations be excited near

$$f = 475 \text{ GHz}$$

The emission line is Doppler broadened to

$$\Delta f = 1500 \text{ MHz}$$

The terminal state for induced transitions has a short enough life for four-level laser operation to be considered. According to (121) the critical inversion density is

$$\Delta n = 5 \times 10^7 \text{ cm}^{-3}$$

If the fluorescence efficiency is assumed near unity and if the power to excite the metastable state 2^3s of helium is taken to correspond to

$$W_{03} = 19{\cdot}81 \text{ eV}$$

the threshold power of fluorescence becomes

$$\frac{P_f}{V} \simeq 2 \times 10^{-3} \frac{\text{W}}{\text{cm}^3}$$

Actually much more power is needed for the indirect excitation within the radio frequency or d.c. discharge, but the above value nevertheless indicates that gas lasers are more easily excited than solid state lasers, and that they will operate cw under quite ordinary conditions.

2.14. Laser structures and their characteristics

In laser structures the active medium must be enclosed by an optical resonator and pumping radiation sources or other excitation sources must be provided. We will only discuss here the most important configurations and their characteristics. These will all be configurations which have worked well under experimental conditions and in addition have found wide use in laser applications.

2.14.1. Gas Lasers

A unique characteristic of gas lasers is the nearly perfect optical homogeneity of the active medium. The most important quality of laser beams, namely their coherence in time and space, may therefore be realized best with gas lasers. A well-proven arrangement is a configuration with mirrors external to the gas discharge. Figure 33 shows such a configuration for a He–Ne laser.

A long cylindrical glass tube is closed on both sides by windows at an angle to the axis corresponding to the Brewster angle. Light waves which travel along the tube axis and are linearly polarized in the plane of incidence at these mirrors will be transmitted without any reflection.

The external mirrors are arranged confocally or have other curvatures depending on the desired spot size for the given mode. The tube diameter is small enough (1 to 3 mm) to keep the density of the neon 1*s*-state low by diffusion to and interaction with the wall. Usually the discharge tube is between 0·1 and 1 m long.

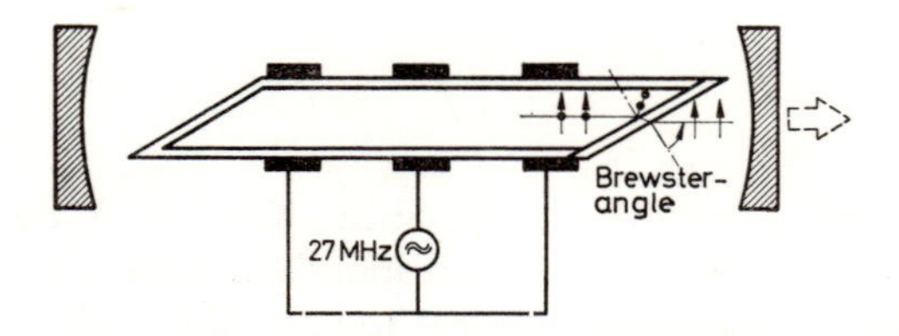

FIG. 33. Gas laser with external mirrors and r.f. excitation

The gas discharge is maintained either by a radio frequency field or by a d.c. electric field. For r.f. excitation, as in Fig. 33, one normally uses 27 MHz. When properly matched the discharge requires 10–50 W of r.f. power.

Many experiments need a uniform and absolutely steady discharge which in this case must be maintained by a d.c. field. Electrodes must then lead into the discharge tube. The d.c. source is required to drive a current of the order of 50 mA at about 1000 V.

A configuration with external mirrors allows an easy exchange and adjustment of mirrors. By changing between frequency selective interference mirrors oscillations may be excited at any one of the different emission lines and the spot size of modes may also be adjusted.

The different emission lines of He–Ne lasers are Doppler broadened from $\Delta f = 300$ MHz at $\lambda = 3{\cdot}39$ μm to $\Delta f = 1500$ MHz at $\lambda = 0{\cdot}6328$ μm. The relatively large optical resonator has many modes of oscillation within this line width. These are not only modes of different transverse orders as shown in Fig. 18. There are in addition for each transverse order a number of different longitudinal orders.

Modes of different longitudinal order have different numbers of half-waves as standing waves between the mirrors. With

$$n\frac{\lambda_n}{2} = L = n\frac{c}{2f_n}$$

for the nth longitudinal order, and

$$(n+1)\frac{\lambda_{n+1}}{2} = L = (n+1)\frac{c}{2f_{n+1}}$$

for the $(n+1)$th longitudinal order the adjacent longitudinal orders are spaced in frequency by

$$\frac{\Delta f}{f} = \frac{f_{n+1}-f_n}{f_n} = \frac{\lambda}{2L}$$

At $L = 1$ m and $\lambda = 0{\cdot}63$ μm we have $\Delta f = 150$ MHz. Within the width $\Delta f = 1500$ MHz of the corresponding emission line we have therefore quite a number of different longitudinal orders (Fig. 34).

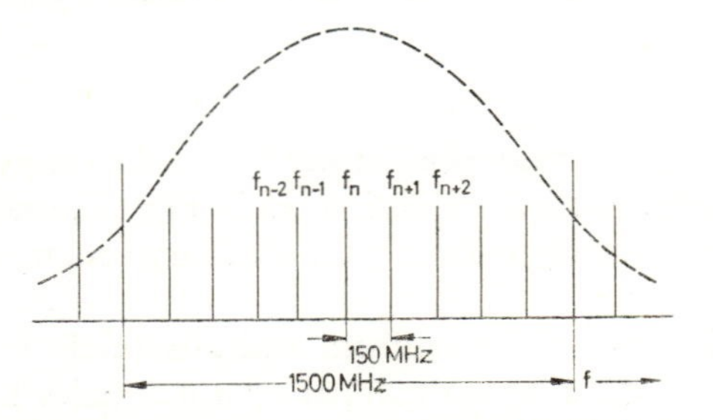

FIG. 34. Resonant frequencies of modes of different longitudinal orders within the Doppler-broadened emission line of a He–Ne laser. The mirrors are spaced by $L = 1$ m

To obtain oscillations at only one single frequency, a mode of lowest transverse order with $m = n = o$ is excited. All modes of higher transverse order are suppressed by limiting the beam cross-section to the spot size of the lowest transverse order with an iris. All higher order modes will then suffer high diffraction losses and cannot be sustained.

To excite only one mode selectively from the different longitudinal orders turns out to be much more difficult. Experiments have been made with several partially transparent mirrors which were spaced to give a highly selective frequency characteristic of the overall reflection.

These experiments and tests of other proposals for single frequency operation have been successful, but they do not all represent practical solutions [29]. It is much simpler to shorten the resonator to a length L which, according to (189), will only leave one single longitudinal order within the range of the Doppler-broadened line. Lasers as short as that will only give a low output power [30].

Two emission lines of the He–Ne mixture are predominantly utilized to generate coherent light:

1. The line at $\lambda = 0{\cdot}6328\ \mu$m because it was the first and for some time the only emission-line of gas lasers in the visible range of the spectrum. It still lends itself very easily to laser operation. Of all gas lasers the He–Ne laser tuned to this wavelength is still the most widely used.
2. The line at $\lambda = 1{\cdot}15\ \mu$m because it shows strong emission just on the border of good sensitivity for infrared photo cathodes. Therefore it can be detected at low level with excellent resolution in time.

Table 2 lists a few characteristics of He–Ne lasers operating at these wavelengths.

TABLE 2.

Characteristics of Gas Lasers

Wavelength	0·6328 μm	1·15 μm
Partial pressures of He–Ne in torr	1:5	1:5
Method of excitation	d.c.	r.f. at 27 MHz
Power requirement	80 W	10–50 W
CW output power	10 mW	40 mW
Frequency fluctuation	2 kHz	2 kHz

2.14.2. Solid-state Lasers

The first successful experiments were made with solid-state lasers having ruby as the active material. Subsequently ruby lasers have found wide use in many different applications. Single crystals of suitably doped ruby are prepared to have the best possible optical uniformity. Ordinarily they are a few millimetres in diameter and a few centimetres long. Both end faces are so ground and polished as to be exactly plane parallel or to have curvatures up to the confocal arrangement. They are then coated to form interference mirrors. Alternatively the optical resonator may also be formed by external mirrors which opens up the possibility of adjustment and exchange of the mirrors.

For excitation, xenon or mercury arc lamps irradiate the crystal from the side to supply the pump energy. The lamps either enclose the crystal helically as in Fig. 35 or, to utilize the radiation energy more efficiently, are placed on the focal axis of an elliptic cylinder (Fig. 36). The ruby

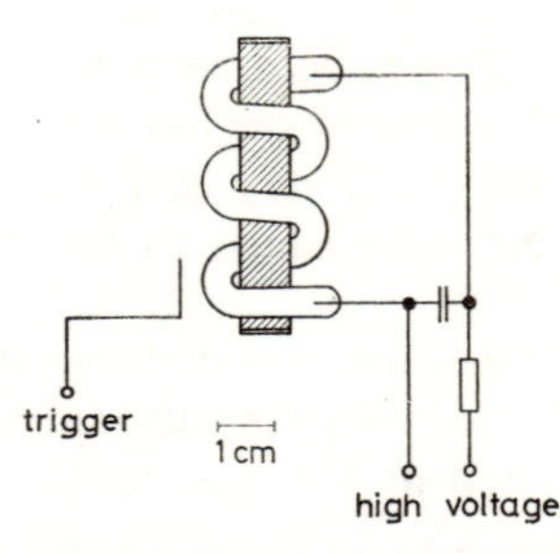

FIG. 35. Ruby laser for pulsed operation with a helical xenon flash tube as the pumping radiation source

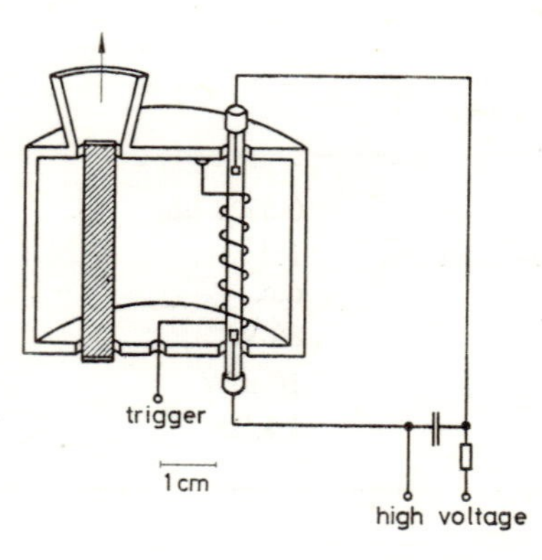

FIG. 36. Ruby laser with an elliptic cylinder for efficient focusing of pumping radiation

crystal is then placed on the other focal axis. The three-level operation of the ruby laser requires so much pumping power that, to avoid undue heating, only pulsed operation is feasible under normal conditions. By discharging condenser banks as in Figs. 35 and 36 the arc lamps generate intense flashes of pumping energy. For cw operation of the ruby laser the output power is low and care must be taken to focus the pumping power efficiently and to dissipate the heat by cooling [31].

Depending on how well the pumping radiation is focused, between 100 to 1000 joules of pumping energy per pulse must be supplied for pulsed operation. The output radiation energy is about 1 joule per pulse. Since pulses will typically last 1 msec, 1 kW is a representative pulse power. During any one pulse, however, the laser beam is not nearly as coherent in time and space as is the case for gas lasers.

The spatial coherence of the laser beam is degraded by optical imperfections of the crystal and a non-uniform inversion density. Even if the end faces were exactly parallel the optical distance between opposite points would depend on location and time. Simultaneously excited oscillations along several narrow paths are observed and these oscillations are not well correlated. The output radiation from different points of one end face is therefore not completely coherent. Also the different paths of oscillation shift their location and spread transversely during one pulse because the inversion density becomes locally depleted.

The time coherence of the laser beam from a particular sub-area of an end face is disturbed mainly by the local depletion of the inversion density. During one pulse of about 1 msec the local amplitude in the laser beam fluctuates. Actually the oscillation very often consists of many individual spikes each lasting for only about 1 μsec. When the density of states is first inverted by the irradiating pump energy and oscillations are excited, the stimulated emission will deplete the population of the upper laser level faster than the pump power can refill this state. Oscillations will than die away and only be re-excited when the pumping radiation has again generated a sufficient inversion of states. In accordance with this explanation of relaxation oscillations the repetition rate of individual spikes within the laser beam increases with pump power.

The local depletion of inversion density not only causes the oscillations to grow and then decay in time, it also shifts small sub-areas of oscillation around and spreads them in time. Taken together these phenomena are quite involved and quite random in time and space.

The lack of spatial and time coherence of the ruby laser broadens the spectrum of its output beam to the order of 10^{10} Hz and spreads the radiation over a beam width of the order of a few minutes of arc. If only low order modes of the optical resonator were excited the beam width according to (168) would be much smaller.

2.15. Q-switching for giant pulses

The relaxation oscillations may be controlled, at least to a limited degree, by the quality factor Q of the optical resonator. In an arrangement with variable Q external mirrors form the optical resonator as shown in Fig. 37. Inside the resonator the light beam passes a polarizer

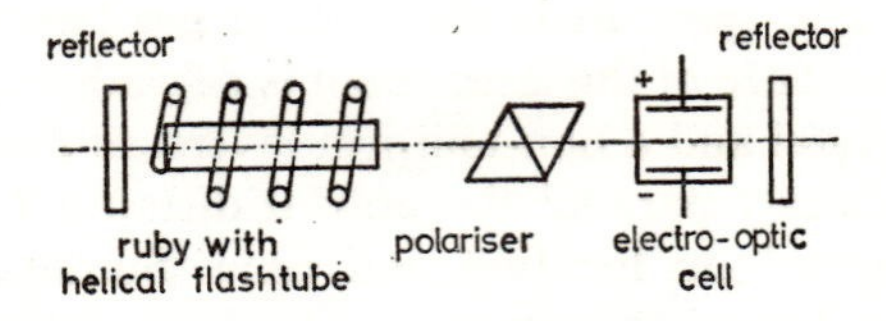

FIG. 37. Electro-optic Q-switching of an optical resonator for the generation of giant pulses

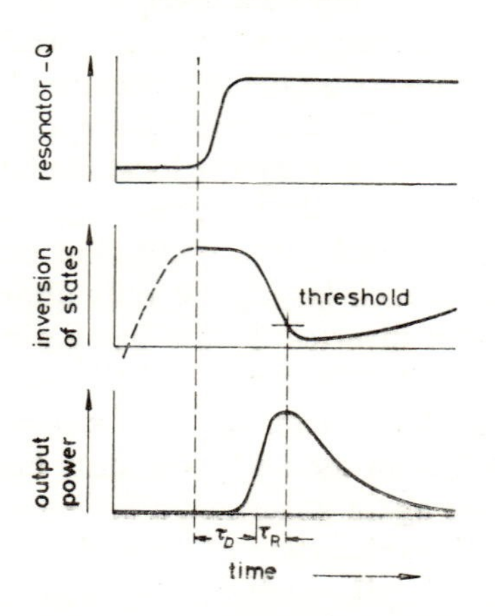

FIG. 38. Time sequence of Q-switching, inversion of states, and output pulse in a Q-switched giant pulse laser

and a Kerr cell. The Kerr cell will be either transparent for one polarization or cause it to be somewhat attenuated through the bias voltage. The Kerr cell voltage thus controls the resonator loss or its Q-factor.

To generate well-defined pulses of high peak power, pumping radiation first inverts the population of laser levels much above threshold while the resonator is still lossy and suppresses any oscillation, then the Kerr cell is suddenly switched into transparency. With the short delay τ_D oscillations will be excited as illustrated by Fig. 38. The inversion of states decreases but the amplitude of the oscillation will continue to grow until after the passage of rise time τ_R when the inversion density has dropped to its normal threshold value. Subsequently the oscillation weakens and the inversion density vanishes entirely due to the continuation of stimulated emission. Such *Q*-switching has been employed to generate pulses of many gigawatts peak power and of a duration shorter than 20 nsec.

The frequency spectrum of giant pulses is not significantly wider than that of the ordinary uncontrolled relaxation oscillations, and the beam width also remains nearly unchanged.

CHAPTER 3

The Maser

MASERS amplify microwaves by induced emission. They are based on the same physical effect as lasers. Many of the basic relations for light amplification in lasers also hold for microwave amplification in masers. The essential difference is in the respective range of operating frequencies. Transitions between states participating in maser action must have transition frequencies in the microwave range. Corresponding energy terms must be much more closely spaced with $\Delta W \simeq 10^{-4}$ eV than in the case of lasers with $\Delta W \simeq 1$ eV. Therefore masers utilize entirely different transitions between states than lasers.

Of the different classes of masers the solid-state maser has found wide practical use for low noise amplification in ultrasensitive receivers for satellite communication systems and radio astronomical observatories. We will therefore limit our discussion to the solid-state maser. In such a maser, using ruby for example, transitions are stimulated between paramagnetic energy levels of chromium ions with which the aluminiumoxide host crystal is doped. Before applying quantum mechanical methods to this process we will illustrate the phenomena with a simple classical model.

3.1. A classical model for paramagnetic resonance

Paramagnetic materials contain a large number of atoms or ions having permanent magnetic dipole moments. In ruby, for example, the Cr^{3+} ions have such magnetic dipole moments, and for solid-state lasers the aluminium oxide of ruby is doped with 10^{20} Cr^{3+} ions per cm^3. The permanent magnetic dipole moment is mainly caused by the spin motion of electrons in the ion. In a d.c. magnetic field these spins will tend to orient themselves parallel to the field lines. While trying to change their orientation to this new direction under the influence of the field forces, the electron spins precess around the d.c. magnetic field.

The angular velocity of this precession is

$$\omega_p = \gamma B_0$$

where B_0 is the magnetic induction of the d.c. field and

$$\gamma = \frac{M}{D}$$

the gyromagnetic ratio of the magnetic dipole moment M of the spin system to its angular momentum D. In solid-state materials the interaction between spins and the lattice of the host crystal leads to damping of the spin precession. A stationary state of thermal equilibrium between spin precession and lattice vibrations develops.

Any alternating magnetic field will also exert a torque on the magnetic moment of the spin. The effect of this torque will be largest for a circular polarization of the alternating magnetic field within the plane normal to the direction of the d.c. field and rotating in the same sense as the spin precession. In addition for maximum effect the a.c. field should have the same frequency ω_p as the spin precession. Under these circumstances the vector of the a.c. magnetic field rotates in synchronism with the spin precession, and its torque acts continuously upon the precession and always in the same sense. The situation corresponds to paramagnetic resonance.

We will start here with an electron spin in a d.c. magnetic field and a spin precession about this field as shown in Fig. 39. If an a.c. magnetic field at resonance with the spin precession as described above is applied, the precession changes. Depending on the relative phase between a.c. field and spin precession the precession amplitude will either increase due to the a.c. field or decrease. During this process the energy of the spin system in the d.c. magnetic field will also change. If the energy grows the difference must be supplied by the a.c. field. In this case the spin system absorbs energy from the field. If the energy decreases the difference adds to the a.c. field energy. In this case the spin system emits energy. Both absorption and emission are then induced by the external a.c. field. The classical model for electron spins in a d.c. magnetic field shows the same phenomena of induced emission and absorption as

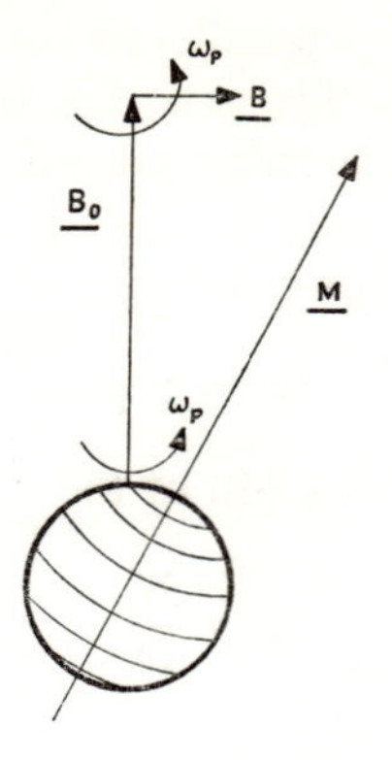

FIG. 39. Classical model of an electron spin precessing about a d.c. field B_0 with a circularly polarized a.c. field B at resonance ω_p

was exhibited by the classical model of electrons oscillating in a potential well.

The energy which is stored in the spin system may be specified and the energy differences from changes in spin precession calculated. For such a calculation we refer to the Hamilton operator for a charged particle in an electromagnetic field. In section 1.7 we have evaluated the interaction component of this Hamiltonian. In a second order approximation which considered the field gradients in the vicinity of the charged particle the operator (46)

$$\overline{H}'_{tM} = -\mathbf{B}\cdot\overline{\mathbf{M}}$$

was found. Here $\mathbf{B}$ is the magnetic induction of the a.c. field and $\overline{\mathbf{M}}$ the magnetic dipole moment operator. $\overline{H}'_{tM}$ has been called the magnetic dipole interaction operator.

The Hamilton operator results from the corresponding Hamilton function by quantization of momentum. This procedure may also be traced backwards and the Hamilton function obtained from its Hamilton operator just by replacing the momentum operators with their corresponding classical momentum functions.

Accordingly

$$H = -\mathbf{B}\cdot\mathbf{M}$$

is the Hamilton function of the magnetic dipole moment **M** in a magnetic field **B**. As such this expression for the Hamilton function also stands for the energy W of the dipole moment in the field. In the case of a d.c. magnetic field $\mathbf{B}_0$ we have

$$W = -\mathbf{B}_0 \cdot \mathbf{M} \tag{190}$$

To obtain the energy of the spin system in a d.c. field we only need to substitute the magnetic moment of the electron spin for **M** in (190).

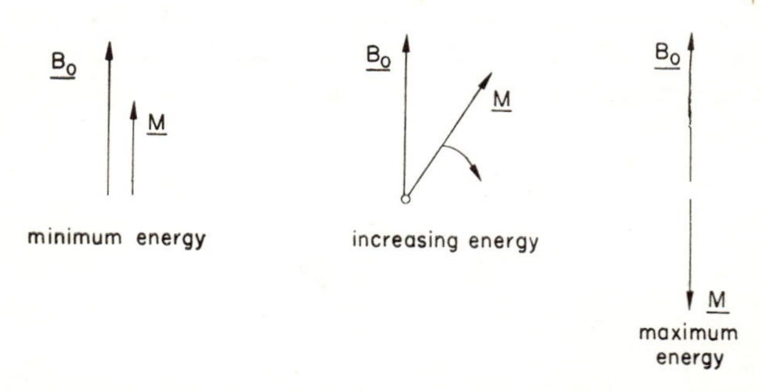

FIG. 40. Orientations of the classical electron spin in a d.c. field

According to (190) the stored energy is lowest when $\mathbf{B}_0$ and **M** have the same direction. There is no precession of **M** in this case. The stored energy grows with the cone angle of precession. It reaches its highest value just when $\mathbf{B}_0$ and **M** have opposing orientations. The spin precession again disappears for this extreme. If an a.c. magnetic field of circular polarization and frequency corresponding to paramagnetic resonance begins to act upon a spin system in the parallel orientation of minimum energy (Fig. 40) a precession is started and the spin energy grows. The electron spin absorbs a.c. field energy. The precession increases up to the cone angle 90° and then continues its change to 180° where the spin system in its antiparallel orientation to the d.c. field has maximum energy. With the interaction continuing the cone angle decreases again from 180° and now the spin system will add energy to the a.c. field. While steadily interacting with the a.c. field the spin system will thus go through periodic cycles during which it first absorbs and then emits energy.

For a large number of spin systems precessing in a d.c. field with random phase and cone angle on the average and for uniform distribution just as much energy will be absorbed by the spins as is emitted from them. Only if a specific state of precession predominates will the a.c. field induce a net energy exchange.

3.2. Quantum mechanical treatment of spin interaction

The classical model for electron spins serves well only in illustrating the phenomena of paramagnetic resonance. For a quantitative description however we must resort to quantum mechanical methods.

According to quantum mechanical considerations the electrons of any atom can remain stationary in discrete states which, in a semi-

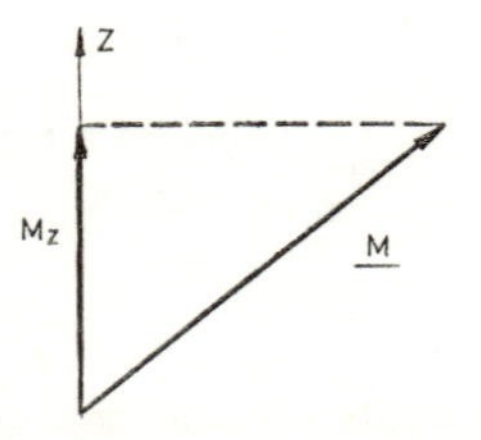

FIG. 41. Vector of magnetic dipole moment and its component M_z parallel to the polar axis for a stationary state of electrons in an atomic field

classical way, are interpreted as electronic orbits around the atomic nucleus. Due to the charge of the orbiting electron each such orbit, and hence each state of the atom, has a specific magnetic moment. The states or orbits of the atom are also quantized with respect to magnitude and direction of their orbital magnetic moments. Therefore an orbital moment cannot be oriented arbitrarily and only certain discrete orientations of orbital moments with respect to each other are possible. In a steady state without external forces the system remains stationary with these discrete orientations of orbital moments. When solving the time-independent Schrödinger equation for stationary states of the electron in the field of the atomic nucleus the different discrete orientations of the orbital momentum are referred to the polar axis of the spherical coordinates for the system. The respective state, and in particular its orbital momentum, are then completely described by the magnitude M of the orbital angular momentum and its component M_z parallel to the polar axis or z-axis of the system (Fig. 41).

Quantization rules similar to those for orbital moments also hold for the electron spin. In a stationary state the electron spin can remain stationary only with certain discrete orientations of the spin vector. For a single electron there are two discrete states with two discrete orientations.

To designate the respective state of the electron spin the description of orbital moments is adopted. The magnitude $|\mathbf{S}|$ of its angular momentum and the component S_z of the angular momentum parallel to the polar axis are specified.

The interaction operator (46) allows a quantum mechanical calculation of the interaction between magnetic fields and electron spins. Eventually such calculations will yield probability values for transitions between different spin states. To evaluate any interaction due to (46) the operator for the magnetic dipole moment of the electron spin must be known. Instead of quantizing the magnetic moment $\mathbf{M}$ it will also serve our purpose to quantize the angular momentum $\mathbf{D}$. With $\mathbf{M} = -\gamma\mathbf{D}$ and γ the gyromagnetic ratio one may be substituted for the other.

In quantum mechanics the angular momentum operator of the electron spins results from commutator rules which, in turn, are postulated in addition to (21) to account correctly for all experimental evidence. For our considerations it will be sufficient to start with the results of this first phase in a quantum mechanical treatment. Instead of postulating commutator rules we will only state their consequences for the spin operators in matrix notation. Since the spin of a single electron has two discrete states the matrices of the spin operator are of second order. The three components of the spin vector have three corresponding matrices for the spin operator. For the three cartesian components the matrices have the following forms:

$$\bar{S}_x = \frac{\hbar}{2}\begin{bmatrix} 0 & 1 \\ 1 & 0 \end{bmatrix}, \quad \bar{S}_y = \frac{\hbar}{2}\begin{bmatrix} 0 & -j \\ j & 0 \end{bmatrix}, \quad \bar{S}_z = \frac{\hbar}{2}\begin{bmatrix} 1 & 0 \\ 0 & -1 \end{bmatrix} \tag{191}$$

Here the z-axis again coincides with the polar axis of the spherical coordinates. We regard (191) as additional postulates which, when properly applied, will account for all experimental facts.

A matrix notation for an operator always refers to a specific system of state functions ψ_n. Here these functions describe stationary states of the system and they are therefore eigenfunctions of the time-independent Schrödinger equation. This is all we need to know about the character and meaning of these functions. For in all subsequent calculations we will only use the matrices (191).

$\bar{S}_z$ is a diagonal matrix. The stationary state functions are therefore also eigenfunctions of $\bar{S}_z$. To show this we multiply

$$\bar{S}_z \psi_n = S_z \psi_n \tag{192}$$

by ψ_n^* and integrate over all space taking into account the orthonormality of ψ_n:

$$\int \psi_m^* \psi_n \, d\tau \equiv \langle m | n \rangle = \delta_{mn} \tag{193}$$

we obtain for the right-hand side of (192) a column matrix $\begin{bmatrix} S_{z1} \\ S_{z2} \end{bmatrix}$ which consists of the eigenvalues of ψ_n. For the left-hand side of (192) we obtain with (191) the column matrix $\dfrac{\hbar}{2}\begin{bmatrix} 1 \\ -1 \end{bmatrix}$. Hence the state functions ψ_n are also eigenfunctions of $\bar{S}_z$ with eigenvalues

$$S_{z1} = \frac{\hbar}{2}, \quad S_{z2} = -\frac{\hbar}{2} \tag{194}$$

As eigenvalues of an operator they are physical observables. Here we are dealing with the *z*-component of the spin operator and the eigenvalues are therefore components of the spin vector parallel to the polar axis for both stationary states ψ_1 and ψ_2.

The relation (193) for orthonormality of the state functions ψ_n used the bracket notation $\langle \cdot | \cdot \rangle$ to denote the integration. The integration has only symbolic character in (193) and therefore such a notation is adequate. Earlier in our discussion such state functions as ϕ and ψ_n in (19) and (24) were still functions of the position coordinates q_i. The state functions of electron spin have no such dependence on coordinates. They are only symbols to specify one of the two possible orientations of the electron spin. Even eqn. (192) is often written with the symbol $|n\rangle$ instead of ψ_n. Representing the latter half of a bracket $\langle n|n\rangle$ this quantity is referred to as "ket". All the ψ_n of a particular equation such as (192) are said to form a ket vector in *n*-dimensional space. The ket in (192) and (193) is said to be an eigen-ket of the operator S_z with eigenvalues S_{zn}.

Furthermore, the symbol $\langle n|$ is called "bra" from the first half of bracket. It designates the complex conjugate state function ψ_n^*. The operation $\langle n|n\rangle$ forms the scalar product of the bra vector with the ket vector.

The abstract notation of sets of state functions by bra and ket vectors has not only proved adequate in the quantum mechanics of the electron spin but it also allows a concise formulation and treatment in many other quantum mechanical problems. It has therefore been adopted quite generally. We will, however, not need this notation in our further discussion because once the operators are replaced by their matrices the state functions will not appear in any subsequent calculations.

According to the postulates of (191) the electron spin assumes one of two states which from (194) have different z-components for their angular momentum. The angular momentum squared comes from (191) as

$$\bar{S}^2 = \bar{S}_x^2+\bar{S}_y^2+\bar{S}_z^2 = \frac{3}{4}\hbar\begin{bmatrix}1 & 0\\0 & 1\end{bmatrix} \tag{195}$$

Here again, as in (191) for $\bar{\mathbf{S}}_z$, we have a diagonal matrix. The stationary state functions ψ_n or their kets $|n\rangle$ are therefore also eigenfunctions and eigen-kets of the operator $\bar{\mathbf{S}}^2$. Furthermore, since the unity matrix appears in (195), the eigenvalues of $\bar{\mathbf{S}}^2$ are identical for both states. Both states, therefore, have the same magnitude of the spin angular momentum

$$|S| = \frac{\sqrt{3}}{2}\hbar \tag{196}$$

The consequences (194), (195) and (196) of postulates (191) may be interpreted physically as follows: The electron spin has two stationary states. Both states have the same magnitude of angular momentum. The z-components of angular momentum are also of equal magnitude but have opposite directions. The x- and y-components are undetermined. The angular momentum vector is therefore directed along a cone with an acute angle in one state and in the other state it is directed along a cone with an obtuse angle (Fig. 42). The cone angles result from

$$\varphi_{1,2} = \text{arc}\cos\frac{S_{z1,2}}{|S|}$$

The orientation of the respective angular momentum on its cone is undetermined. All these orientations are equally probable. In a statistical sense the spin vector is uniformly spread over the whole circum-

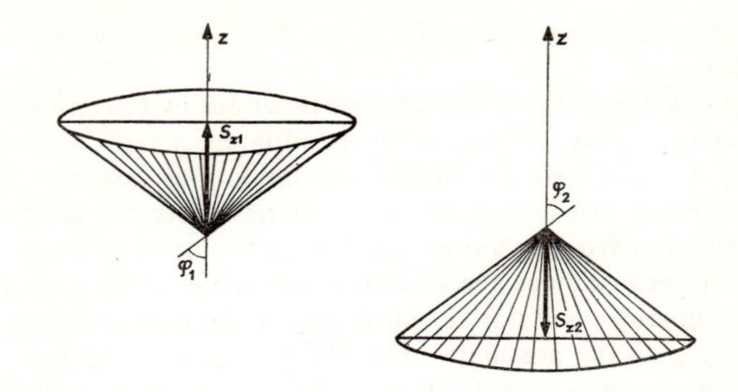

FIG. 42. Cones for the orientation of angular momentum vectors in both states of the single electron spin

ference of the respective cones. Only the component of the spin vector parallel to the cone axis may be observed physically. All other components cancel in their time average. One state of the electron spin has an effective component $S_{z1} = \hbar/2$ of the spin vector and the other state $S_{z2} = -\hbar/2$. Both states of the isolated electron spin differ only in the z-components of their angular momentum.

We refer here to a coordinate system which may still be oriented in any arbitrary direction. If the electron spin is not exposed to a magnetic field both spin states indeed have equal energy levels; they are degenerate. Any linear combination of degenerate eigenfunctions will again be an eigenfunction of the system. In accordance with the arbitrary orientation of the coordinate system both states of the electron spin are therefore still undetermined. Only with a magnetic field present will both states have different energy levels according to (190). The magnetic field removes the degeneracy and both states of the electron spin may be determined.

The magnetic fields which act upon electron spins may be internal fields of the atom or the crystal lattice as they originate from magnetic dipoles of orbital moments or from spins of other electrons. Electron spins may also interact with external fields. To calculate the spin energy in a field from (190) the magnetic dipole moment of the electron spin must be substituted for **M**. In case of a particle of mass m and charge $-e$ the orbital angular momentum **D** and magnetic dipole moment **M** are related by

$$\mathbf{M} = -\gamma \mathbf{D}$$

With the gyromagnetic ratio $\gamma = e/2m$.

The general relativistic theory of the electron and experimental evidence requires the corresponding relation for the electron spin to be different. In case of electron spins we have the operator equation

$$\mathbf{M} = -g_L \cdot \gamma \cdot \mathbf{S} \tag{197}$$

where g_L is called the Landé factor. For the isolated electron $g_L = 2{\cdot}0023$ which, for all practical purposes, may be approximated by $g_L = 2$. Hence the magnetic dipole moment is twice as large as is to be expected from the classical relation between angular and magnetic momenta.

The negative charge of the electron causes the magnetic moment to have an orientation opposite to that of the angular momentum. According to (190) the energy level in a d.c. field $\mathbf{B}_0$ follows from the scalar product of $\mathbf{B}_0$ and the vector $\mathbf{M}{\cdot}\mathbf{M}$ in turn is determined through $\mathbf{S}$ from (191) and has sharply defined components only in the direction of z, since only the matrix of the z-component for the operator $\bar{\mathbf{S}}$ is diagonal. To obtain the required stationary solution we must therefore choose the z-direction parallel to $\mathbf{B}_0$. With this choice of z-orientation the uncertain components S_x and S_y do not enter the product $\mathbf{B}_0 \cdot \mathbf{M}$ and definite energy levels result for both states of the electron spins

$$W_1 = -g_L \gamma S_{z1} B_0 = -\beta B_0, \quad W_2 = -g_L \gamma S_{z2} B_0 = \beta B_0$$

The quantity $\beta = \hbar\gamma = e\hbar/2m$ is called the Bohr magneton. The energy difference between the levels again defines a transition frequency

$$\omega_{12} = \frac{W_2 - W_1}{\hbar} = 2\gamma B_0 \tag{198}$$

For the isolated electron spin we have a gyromagnetic ratio:

$$\gamma_s = M_s/S_z = g_L \gamma \simeq 2\gamma$$

The transition frequency ω_{12} therefore just turns out to be the angular frequency ω_p as for the precession of an electron spin around the magnetic field $\mathbf{B}_0$ in the classical model. With a mass m and charge $-e$ for the electron the gyromagnetic ratio has the following value

$$\gamma = 8{\cdot}79 \times 10^{10} \frac{\text{m}^2}{\text{Vsec}^2}$$

Transition frequencies in the microwave range $f_{12} = 0{\cdot}3 \ldots\ 30$ GHz will therefore require d.c. magnetic fields of the order $\mathbf{B}_0 = 10^2 \ldots 10^4$ G· Such field strengths can be readily produced using permanent magnets.

Transition between states of the electron spin caused by a.c. magnetic fields are governed by the Hamilton operator (46) for magnetic dipole interactions. For our present purpose we can calculate the power which spins and fields exchange during such transitions when we proceed from eqn. (84).

We need only replace the matrix element $\mathbf{E}_0 \cdot \mathbf{P}_{12}$ of electric dipole interaction by the matrix element $\mathbf{B} \cdot \mathbf{M}_{12}$ of magnetic dipole interaction in (84) to obtain the power conversion between spins and fields. Here $\mathbf{B}$ designates the complex magnetic induction phasor of an a.c. magnetic field. The power exchanged during induced transitions between states 1 and 2 is

$$P = \frac{\pi}{\hbar} |\mathbf{B} \cdot \mathbf{M}_{12}|^2 (N_1 - N_2) f g(f) \tag{199}$$

where N_1, N_2 and $g(f)$ still have the same meaning as before. N_1 denotes the number of spins in state 1, N_2 the number of spins in state 2, and $g(f)$ is the normalized line shape. The emission line can again be broadened homogeneously and inhomogeneously. Line broadening due to limited lifetime of spin states shapes the line according to the Lorentz curve. Line broadening by scattering of energy levels results in a Gaussian curve. Lifetime in states is not only limited by spontaneous transitions but also due to relaxation through interaction between spins and the lattice of the host crystal. Scattering of energy levels in a solid-state maser is caused by the surrounding crystal lattice which can have variations in its lattice field from micro-system to micro-system.

The matrix element M_{12} in (199) of the magnetic dipole interaction operator has a vector character; its Cartesian components in (x, y, z) are the elements (1, 2) of the corresponding matrices. Following the customary notation we write

$$\mathbf{M}_{12} = -g_L \beta \boldsymbol{\sigma}$$

where $\boldsymbol{\sigma}$ then designates a dimensionless vector. Its Cartesian components, obtained from the matrix elements (1, 2) of the pi operator, are

$$\sigma_x = \frac{1}{2}, \quad \sigma_y = -\frac{j}{2}, \quad \sigma_z = 0 \tag{200}$$

For the scalar product in (199) we have, in this notation,

$$|\mathbf{B}\cdot\mathbf{M}_{12}|^2 = g_L^2\beta^2\mu_0^2(\mathbf{H}^+\boldsymbol{\sigma}\boldsymbol{\sigma}^+\mathbf{H}) \tag{201}$$

where $\boldsymbol{\sigma}\cdot\boldsymbol{\sigma}^+$ is a dyadic with elements obtained from

$$\boldsymbol{\sigma}\cdot\boldsymbol{\sigma}^+ = \begin{bmatrix}\sigma_x\\ \sigma_y\\ \sigma_z\end{bmatrix}[\sigma_x^*\sigma_y^*\sigma_z^*] = \begin{bmatrix}\sigma_x\sigma_x^* & \sigma_x\sigma_y^* & \sigma_x\sigma_z^*\\ \sigma_y\sigma_x^* & \sigma_y\sigma_y^* & \sigma_y\sigma_z^*\\ \sigma_z\sigma_x^* & \sigma_z\sigma_y^* & \sigma_z\sigma_z^*\end{bmatrix} \tag{202}$$

In the case of (200) for the isolated electron spin this dyadic reduces to

$$\boldsymbol{\sigma}\cdot\boldsymbol{\sigma}^+ = \frac{1}{4}\begin{bmatrix}1 & j & 0\\ -j & 1 & 0\\ 0 & 0 & 0\end{bmatrix} \tag{203}$$

When we now substitute (201) into (199) and divide the stimulated power conversion by the volume, the power density from the induced spin transitions is obtained

$$\frac{P}{V} = \frac{\pi}{\hbar}(n_1-n_2)fg(f)g_L^2\beta^2\mu_0^2(\mathbf{H}^+\boldsymbol{\sigma}\cdot\boldsymbol{\sigma}^+\mathbf{H}) \tag{204}$$

If $n_1-n_2>0$ the power density is positive and we have induced absorption, when $n_1-n_2<0$ it is negative and we have induced emission.

A corresponding relation between converted power density and magnetic field follows from the conservation of complex energy for steady state fields of harmonic time dependence. From the conservation of energy

$$\frac{P}{V} = -\omega\,\mathrm{Im}(\mathbf{H}^*\cdot\mathbf{B}) = \omega\mu_0(\mathbf{H}^+\chi_m''\mathbf{H}) \tag{205}$$

is the power per unit volume which a material with a complex tensor of magnetic susceptibility $\boldsymbol{\chi}_m$ and a loss component

$$\chi_m'' = \frac{1}{2j}(\chi_m-\chi_m^+)$$

absorbs from a magnetic field $\mathbf{H}$.

Comparing (204) and (205) the induced absorption or emission may be accounted for by the imaginary component

$$\chi''_m = \frac{1}{2\hbar}(n_1 - n_2)g(f)g_L^2\beta^2\mu_0\sigma\sigma^+ \tag{206}$$

of a complex tensor of magnetic susceptibility.

In case of an isolated electron spin we have

$$\chi''_m = \frac{1}{8\hbar}(n_1 - n_2)g(f)g_L^2\beta^2\mu_0 \begin{bmatrix} 1 & j & 0 \\ -j & 1 & 0 \\ 0 & 0 & 0 \end{bmatrix} \tag{207}$$

From the classical model for the electron spin we expect the interaction to become largest for a magnetic field with circular polarization in the (x, y) plane. Such a polarization has the following components

$$\mathbf{H} = \frac{H}{\sqrt{2}} \pm (\mathbf{u}_x \mp j\mathbf{u}_y) \tag{208}$$

where with the upper signs it describes a positive sense of rotation of the field vector and with the lower signs a negative sense of rotation. For the governing factor in (203) we obtain by substituting from (208)

$$(\mathbf{H}^+\sigma\sigma^+\mathbf{H}) = \begin{cases} \dfrac{H^2}{2} & \text{for + polarization} \\ 0 & \text{for − polarization} \end{cases} \tag{209}$$

The circular polarization with positive sense of rotation has the largest effect and induces maximum power conversion. No power is converted when the field rotates in a negative sense.

Since both the last row and the last column of the matrices in (203) and (207) vanish, no power is converted during the interaction with a magnetic field of linear polarization in the z-direction. To calculate the power conversion for an a.c. magnetic field of a more general polarization it should be separated into these three polarization components. Only the component with positive circular polarization in a plane perpendicular to the d.c. field is then extracted. All other components will have no effect. The same characteristics were also found for the classical electron spin model.

3.3. Paramagnetic energy levels and induced transitions

For solid-state masers the transitions induced by microwaves are between different electron spin states of the ions in a crystal lattice. Especially the metallic ions from the transition group have a permanent magnetic dipole moment caused mainly by electron spins when placed in a suitable crystal lattice. Among the paramagnetic substances in this general classification ruby with its Cr^{3+} ions in aluminum oxide has proved to be good maser material.

Our preceding analysis of the isolated electron spin interacting with a magnetic field will form a suitable basis for the discussion of paramagnetic energy levels and induced transitions between them. The electron spin energy levels in paramagnetic substances depend on external d.c. magnetic fields in a much more involved fashion than do the free electron spin levels. The interaction between spin states and a.c. magnetic fields is also further complicated by the effects of the ion and the surrounding crystal lattice.

In the first place the magnetic dipole moment of the ion is not just caused by one electron spin. There are rather a number of electrons with spins contributing to the magnetic dipole moment. In addition the orbital momenta contribute to the total dipole moment.

The strong fields of the surrounding crystal lattice quench the orbital momenta to a large degree, however, while they leave the spins more or less unaffected. As a consequence the Landé factor relating angular momentum to magnetic momentum is only a little less than two. Orbital momenta alone would cause the Landé factor to be unity while electron spins alone would make $g_L = 2$. With g_L nearly equal to 2 the paramagnetic character therefore comes mainly from the electron spins.

The magnetic characteristics of any atom are determined by the electrons of the outer shell which are not engaged in chemical bonding or not exchanged with the crystal lattice. An atom will only show a permanent magnetic moment when the outer shell is only partially filled according to Pauli's exclusion principle.

The Cr^{3+} ion with all orbital energies at ground level has three electrons in the outer shell. Pauli's principle would allow ten electrons in this shell. The spins of these three outer electrons can combine in four different orientations as shown in Fig. 43. We therefore have four spin states with four different energy levels in a d.c. magnetic field. But even without any external magnetic field the spin states do not degenerate entirely to one level. Due to the internal fields of Cr^{3+} in ruby one pair of the four spin states has an energy level differing from the other pair. The change in all four energy levels due to an external magnetic field has been plotted in Fig. 44 for a representative orientation of the d.c. field with respect to the axis of symmetry of the ruby

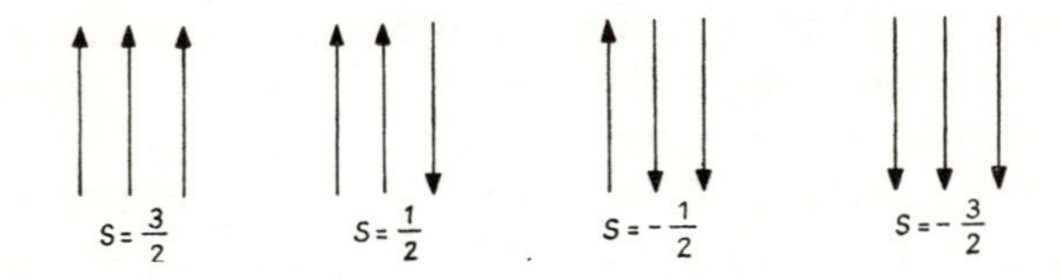

FIG. 43. Spin states of the three electrons in the outer shell of Cr^{3+} ions

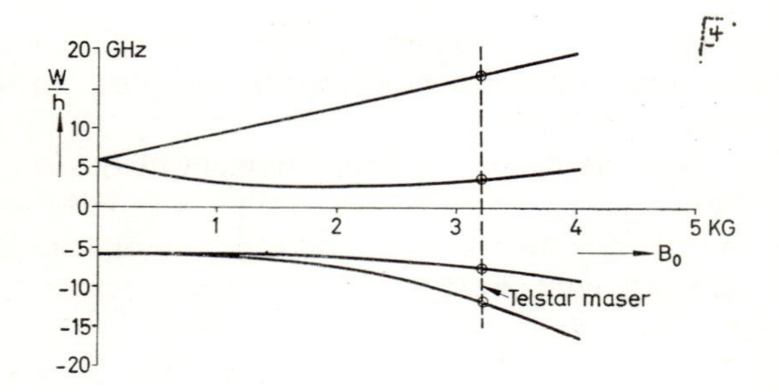

FIG. 44. Energy levels for the four spin states of Cr^{3+} in ruby with d.c. field B_0 normal to the crystal axis

host crystal. In contrast to the energy levels of free electron spins these levels do not depend linearly on field strength any more and some of them do not even change monotonically with field strength.

The splitting of these degenerate states into different energy levels due to external magnetic fields is called the *Zeeman effect*. The different energy levels which arise are called *Zeeman levels* or *paramagnetic levels*.

A.c. magnetic fields of suitable frequency and polarization will induce transitions not only between spin states of free electrons but also between spin states of the ions of the transition group in a crystal lattice. Due to the internal crystal fields, however, the spin operator matrix elements have a more complex structure than is the case for the free electron spin in (195). The components of the σ-vector in (200) will now be complex. In addition they depend on the magnitude of the d.c. magnetic field and, because of crystal anisotropy, also on the direction of the d.c. field with respect to the crystal axis.

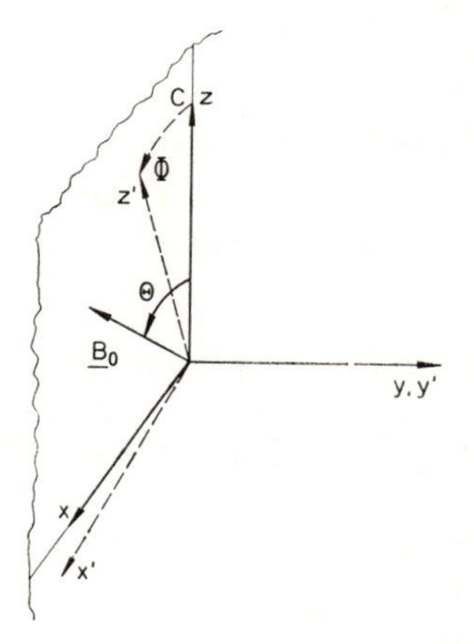

FIG. 45. Orientation of the d.c. magnetic field with respect to the axis of symmetry and coordinate systems for axially symmetrical crystals

The situation is somewhat simplified when, as is the case for ruby, the ion Cr^{3+} has just four spin states and the host crystal Al_2O_3 has axial symmetry. For such a case Fig. 45 shows a suitable orientation of the coordinate system. The z-axis coincides with the axis of symmetry, e.g. the c-axis of the crystal, and the d.c. magnetic field lies parallel to the $(x-z)$ plane. For this coordinate system the σ-vector for any of the six transitions between the four spin states may be written as follows:

$$\boldsymbol{\sigma}_{ij} = \tfrac{1}{2}\left[\mathbf{u}_x\alpha_{ij} + j\mathbf{u}_y\beta_{ij} + \mathbf{u}_z\gamma_{ij}\right] \tag{210}$$

Under the present conditions of only four spin states in an axially symmetric host crystal the quantities α_{ij}, β_{ij}, and γ_{ij} are all real but they depend on the magnitude B_0 and the direction of the d.c. field and have different values for each of the transitions between the spin states. For Cr^{3+} in ruby extensive compilation of the numerical values for these quantities exists [14].

The loss or gain component of the susceptibility tensor follows from $\boldsymbol{\sigma}\boldsymbol{\sigma}^{+}$. This tensor, using (210), is obtained from (202), i.e.

$$\boldsymbol{\sigma}\,\boldsymbol{\sigma}^{+} = \frac{1}{4}\begin{bmatrix} \alpha^2 & -j\alpha\beta & \alpha\gamma \\ j\alpha\beta & \beta^2 & j\beta\gamma \\ \alpha\gamma & -j\beta\gamma & \gamma^2 \end{bmatrix} \tag{211}$$

A rotation of coordinates will reduce (211) to a simpler form. If the (xyz)-system is rotated around the y-axis by an angle

$$\phi = \text{arc tan}\,\frac{\gamma}{\alpha} \tag{212}$$

then in the rotated system $(x'\,y'\,z')$ the $\boldsymbol{\sigma}$-vector takes the form

$$\boldsymbol{\sigma} = \tfrac{1}{2}\left[\mathbf{u}_{x'}\sqrt{\alpha^2+\gamma^2}+j\mathbf{u}_{y'}\,\beta\right]$$

As for the free electron spin there is no $\boldsymbol{\sigma}$-vector z'-component in the $(x'\,y'\,z')$ system. Furthermore in the $(x'\,y'\,z')$ coordinates the dyadic product $\boldsymbol{\sigma}\,\boldsymbol{\sigma}^{+}$ reduces to

$$\boldsymbol{\sigma}\boldsymbol{\sigma}^{+} = \frac{1}{4}\begin{bmatrix} \alpha^2+\gamma^2 & -j\beta\sqrt{\alpha^2+\gamma^2} & 0 \\ j\beta\sqrt{\alpha^2+\gamma^2} & \beta^2 & 0 \\ 0 & 0 & 0 \end{bmatrix} \tag{213}$$

All elements in the last row and column vanish. These elements were also zero in the corresponding matrix for the isolated electron spin. Thus the component of the a.c. magnetic field in the z'-direction will not induce any transitions. Interaction occurs only through the x'- and y'-components. This interaction will be most effective when the a.c. magnetic field has an orientation parallel to $\boldsymbol{\sigma}$. From (213) the field orientation for maximum interaction requires the field phasor to have the following form:

$$\mathbf{H} = H\,\frac{\mathbf{u}_{x'}\sqrt{\alpha^2+\gamma^2}+j\mathbf{u}_{y'}\beta}{\sqrt{\alpha^2+\beta^2+\gamma^2}}$$

This optimum field is elliptically polarized. The governing factor for power conversion becomes

$$\mathbf{H}^{+}\boldsymbol{\sigma}\boldsymbol{\sigma}^{+}\mathbf{H} = \tfrac{1}{4}\,H^2(\alpha^2+\beta^2+\gamma^2)$$

with the optimum polarization. It represents the maximum power conversion. If the field is elliptically polarized according to either one of the following forms

$$\mathbf{H} = H \pm \frac{\mathbf{u}_{x'}\beta \pm j\mathbf{u}_{y'}\sqrt{\alpha^2+\gamma^2}}{\sqrt{\alpha^2+\beta^2+\gamma^2}} \tag{214}$$

the governing factor for power conversion becomes

$$(\mathbf{H}^{+}\sigma\sigma^{+}\mathbf{H}) = \begin{cases} H_{+}^{2}\dfrac{(\alpha^2+\gamma^2)\beta^2}{\alpha^2+\beta^2+\gamma^2} & \text{for + polarization} \\ 0 & \text{for − polarization} \end{cases} \tag{215}$$

For this polarization the medium has a non-reciprocal characteristic similar to the isolated electron spins. The non-reciprocal character of these paramagnetic crystals is most pronounced for such an elliptical polarization while in the case of isolated electrons it was stronges tfor a circular polarization.

For an a.c. magnetic field with linear polarization we have

$$(\mathbf{H}^{+}\sigma\sigma^{+}\mathbf{H}) = \begin{cases} H_{x'}^{2}\,\dfrac{\alpha^2+\beta^2}{4} & \text{for } x' \text{ polarization} \\ H_{y'}^{2}\,\dfrac{\beta^2}{4} & \text{for } y' \text{ polarization} \end{cases} \tag{216}$$

As a representative example Fig. 46 shows $\frac{1}{4}(\alpha^2+\beta^2)$ as a function of the orientation θ of the d.c. field for all six transitions between the four spin states.

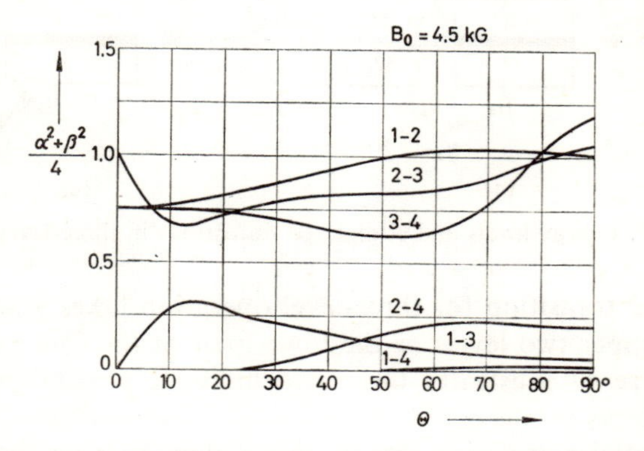

FIG. 46. Factor $(\alpha^2+\beta^2)/4$ for transitions which are induced by a field linearly polarized in x'-direction [14]

The curves of Fig. 46 show that in general transitions between neighbouring states are more easily stimulated than the other transitions where two or even three of the electrons change their spin states simultaneously. Numerical tables for α, β und γ in addition show that quite generally for transitions between neighbouring levels the z'-axis is oriented nearly parallel to the d.c. magnetic field and according to (214) and (215) that for maximum non-reciprocity the a.c. field has nearly circular polarization. For these reasons the isolated electron spin is an excellent model to describe the behaviour of ruby and similar materials at paramagnetic resonance.

3.4. Three-level maser

Of the four different paramagnetic energy levels in ruby normally only three participate in stimulated transitions for maser operation. It is therefore called a three-level maser but with a connotation different from the three-level operation in lasers as is obvious from Fig. 47. This shows two different examples of energy levels for three-level masers.

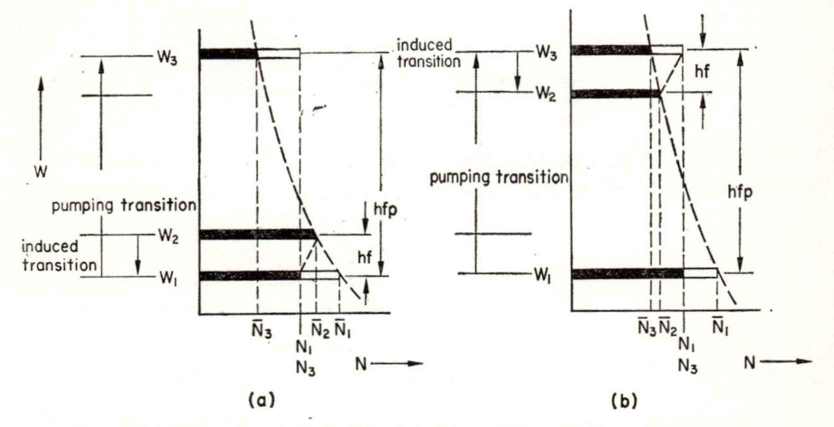

FIG. 47. Energy levels and stimulated transitions in three-level masers

The signal transition for three-level operation takes place either between the upper two levels or the lower pair of the four paramagnetic levels. Figure 47 illustrates these two different possibilities with diagrams of energy versus population of levels. The broken curve indicates the exponential population distribution following from the Boltzmann statistics for thermal equilibrium.

For both modes of operation the pumping transition is between the lowest and uppermost level. The population of these two levels when pumped to complete saturation has also been indicated in Fig. 47. Sufficiently high pumping power will stimulate many more transitions from level 1 to level 3 and vice versa than happen spontaneously or by relaxation. Under these circumstances both these levels have equal population while the population of the intermediate levels is almost that given by the Boltzmann distribution. Depending on temperature and the location of the intermediate level we will have a population inversion arising either between levels 1 and 2 or between 2 and 3.

Indicating population numbers at thermal equilibrium by a raised index (e) we have, according to Boltzmann statistics,

$$\frac{\bar{N}_2^{(e)}}{\bar{N}_1^{(e)}} = e^{-\frac{W_2-W_1}{kT}}, \qquad \frac{\bar{N}_3^{(e)}}{\bar{N}_1^{(e)}} = e^{-\frac{W_3-W_1}{kT}}$$

Pumping to complete saturation results in

$$N_1 = N_3 = \frac{\bar{N}_1^{(e)}+\bar{N}_3^{(e)}}{2} = \frac{\bar{N}_1^{(e)}}{2}\left(1+e^{-\frac{W_3-W_1}{kT}}\right)$$

To calculate the population N_2 of the intermediate levels during maser operation we must take into account all transitions which are either stimulated by signals and thermal radiation or due to spin lattice relaxation. In addition we must account for all transitions which take place spontaneously. All transitions which are stimulated by thermal radiation and spin lattice relaxation and which occur spontaneously are summarily called thermal transitions. Spontaneous transitions will not only emit photons which add to the thermal radiation but, during the spin lattice relaxation, also add phonons to the thermal lattice vibration. Thermal transition in solid-state material with transition frequencies in the microwave range consist mainly of transitions due to spin lattice relaxation. These thermal spin lattice transitions dominate thermal radiation transitions by a few orders of magnitude.

During interaction with a signal in the frequency range $f_{12} = \frac{W_2 - W_1}{h}$ the rate of change with time of the intermadiate level population N_2 follows from

$$\frac{dN_2}{dt} = (s_{12}+w_{12})N_1-(s_{21}+s_{23}+w_{21})N_2+s_{32}N_3 \tag{217}$$

Here w_{ij} again denotes the transition rate which the signal induces per micro-system from level i to j. s_{ij} denotes the corresponding rate of thermal transitions. For signal induced transitions the rate is equal in both directions $w_{12} = w_{21}$, while for thermal transitions the relation

$$s_{ij} \cdot N_i^{(e)} = s_{ij} \cdot N_j^{(e)}$$

follows from conditions at thermal equilibrium. With the Boltzmann distribution for equilibrium population the thermal transition ratio is

$$\frac{s_{ij}}{s_{ji}} = e^{-\frac{W_j - W_i}{kT}} = e^{-\frac{hf_{ij}}{kT}} \tag{218}$$

If we restrict our considerations to frequencies in the microwave range ($f \leqslant 20$ GHz) and temperatures above the boiling point of helium ($T > 4°$K) we always have

$$\frac{hf}{kT} < \frac{1}{4} \tag{219}$$

Under these conditions we may approximate (218) by

$$\frac{s_{ij}}{s_{ji}} \simeq 1 - \frac{hf_{ij}}{kT} \tag{220}$$

Also under these conditions the various paramagnetic levels will be populated nearly equally. With $N = N_1 + N_2 + N_3$ we then have approximately

$$N_i \simeq \tfrac{1}{3} N \tag{221}$$

Substituting from (220) and (221) into (217) results in

$$\frac{dN_2}{dt} = (s_{21} + w_{21})N_1 - (s_{21} + s_{32} + w_{21})N_2 + s_{32}N_3 - \frac{hN}{3kT}(s_{21}f_{21} - s_{32}f_{32})$$

The approximation (221) has been utilized here only for the last term which because of (219) is quite small and therefore less significant.

For steady state the time rate of change $dN_2/dt = 0$ must vanish. If, in addition, we pump to full saturation with $N_1 = N_3$, the population difference of the intermediate level follows from

$$N_2 - N_1 = N_2 \quad N_3 = \frac{hN}{3kT} \frac{s_{32}f_{32} - s_{21}f_{21}}{s_{21} + s_{32} + w_{21}} \tag{222}$$

To obtain population inversion between levels 1 and 2 we must have

$$\frac{s_{32}}{s_{21}} > \frac{f_{21}}{f_{32}} \tag{223}$$

The alternative three-level operation is amplification of a signal in the frequency range $f_{23} = \frac{W_3 - W_2}{h}$ by stimulated transitions between levels 3 and 2. In this case with a signal induced transition rate $w_{32} = w_{23}$ the population difference follows from

$$N_3 - N_2 = N_1 - N_2 = \frac{hN}{3kT} \frac{s_{21}f_{21} - s_{32}f_{32}}{s_{21} + s_{32} + w_{32}} \tag{224}$$

To obtain population inversion between levels 3 and 2 we must have

$$\frac{s_{21}}{s_{32}} > \frac{f_{32}}{f_{21}} \tag{225}$$

Often it is more convenient to specify not the thermal transition rate s but the corresponding relaxation time

$$\tau = \frac{1}{s}$$

With these quantities condition (223) for population inversion appears as

$$\frac{f_{32}}{f_{21}} > \frac{\tau_{32}}{\tau_{21}}$$

while (225) takes the form

$$\frac{f_{21}}{f_{32}} > \frac{\tau_{21}}{\tau_{32}}$$

When both relaxation times are nearly equal the population will be inverted between 1 and 2 when $_2$ is closer to 1 than to 3 otherwise the population will be inverted between 2 and 3. In some solid-state maser materials the respective relaxation times can differ considerably however.

By generalizing the notation in (222) and (224) both these equations may be combined into one. Let ΔN denote the inversion of states, f the transition frequency in the range of signal frequencies, and f_p the pump frequency. Furthermore let s denote the rate of thermal transitions at the signal frequency, s' the thermal rate at the other transition and w the induced transition rate. Using this notation we obtain from (222) and (224)

$$\Delta N = \frac{hN}{3kT}\,\frac{s'(f_p - f) - sf}{s' + s + w} \tag{226}$$

The power emitted from stimulated transitions is proportional to both the inversion of states ΔN and the induced transition rate w:

$$P = hfw\Delta N = \frac{h^2 Nf}{3kT}\,\frac{s'(f_p - f) - sf}{s' + s + w}\,w \tag{227}$$

w, in turn, is directly proportional to the stimulating power density. Therefore, and according to (227), the emitted power will only depend linearly on the driving power as long as $w \ll s' + s$. As soon as induced transitions become as numerous as thermal transitions the active medium goes into saturation. The gain decreases and eventually for $w \gg s' + s$ the emitted power will be independent of the driving power at a saturation value of

$$P = \frac{h^2 Nf}{3kT}\,[s'(f_p - f) - sf] \tag{228}$$

To obtain large emission according to (227) and, as a consequence of it, a high gain, the maser material should have a large concentration of active micro-systems in the host crystal. Furthermore, the temperature should be very low. To realize any useful gain at all requires cryogenic cooling. For efficient operation liquid helium at or below 4·2°K is used as a cooling agent and for high gain the spacing between frequencies $f^p - f$ should be as large as possible, i.e. the pumping transition should have a much larger difference in energy levels than the signal transition.

The dependence of induced emission and gain on N requires one further comment. According to (226) the saturated inversion of states ΔN cannot be increased indefinitely by just increasing the number N of active micro-systems. Too high a concentration of paramagnetic ions means a small spacing and stronger interaction between them. The ratio of relaxation times $\tau/\tau' = s'/s$ will be decreased due to such interaction and population inversion as well as gain will also be lower.

3.5. Travelling wave maser

To utilize a maser material for amplification certain conditions must be fulfilled. First it must be arranged in a suitable microwave circuit so that a pumping oscillation at frequency f_p has the proper magnetic field polarization to invert the population of maser states almost into saturation. Furthermore, the signal frequency f in the microwave circuit must have a magnetic field of a polarization and strength which will interact efficiently with the maser transitions. A circuit separation must be provided between the input signal at this frequency and the amplified output signal. In addition the maser material must be magnetized in a d.c. bias field and cooled cryogenically.

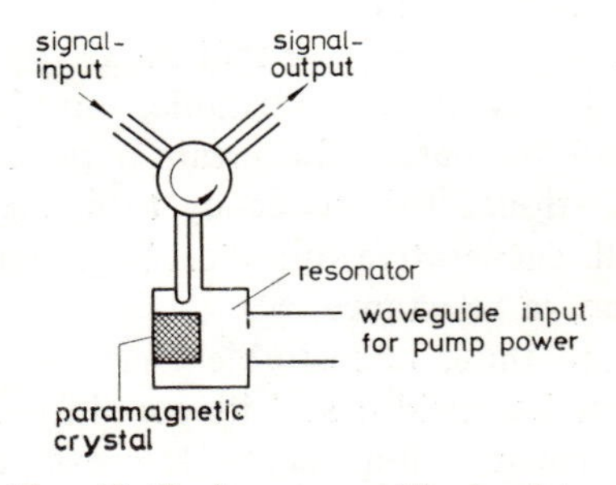

FIG. 48. Cavity-maser with circulator

All these conditions are met, if not in an entirely perfect manner, with *cavity resonators.* In particular, the resonance "step-up" of fields in such a cavity serves to enhance the interaction of pump and signal oscillations. Provided the quality factors of the modes at pumping and signal frequencies are sufficiently high the respective magnetic fields will be strong even for low excitation. They will then stimulate many transitions. Such a cavity maser will normally have only one port for the signal input and output. The amplification of the maser material will cause the input impedance at this port to have a negative resistive component in the signal frequency range and the reflection coefficient magnitude will become larger than unity. At the signal port of the cavity maser an incident signal wave will be reflected with an amplitude larger than the incident amplitude. Input and output waves may be separated by a circulator. The amplifier arrangement is shown in Fig. 48 and the directions of signal flow are also indicated.

Because the magnetic field is enhanced at resonance the gain of a cavity maser increases when the quality factor is improved. A higher Q, however, will also narrow the resonance width. The Q factors required for net gain in a cavity maser are so high that the resonance width is smaller than the width of the emission line. This resonance width therefore limits the bandwidth of a cavity maser. With increasing gain the bandwidth decreases, while the gain bandwidth product remains nearly constant.

Low noise preamplifier stages of sensitive receivers require that the maser has a certain minimum gain, otherwise the sensitivity would be degraded by the higher noise introduced by subsequent amplifier stages. For this minimum gain the bandwidth is quite often too narrow for specific applications. The cavity maser with adequate gain never utilizes the full width of the emission line.

As a further disadvantage we have that the cavity maser is only conditionally stable and may become unstable, especially when tuned to high gain where small fluctuations in certain parameter values may lead to self-excited oscillations. This tendency to excite unwanted oscillations is typical for all regenerative amplifiers based on the reflection from a differentially negative resistance.

The effects of both these undesirable characteristics of the cavity maser, i.e. its narrow bandwidth and its conditional stability, may be reduced by special circuit arrangements. However these arrangements are fairly complex and, therefore, also costly [32, 33].

More efficient and reliable operation may be obtained with a travelling wave maser. In this arrangement the maser material continuously loads a waveguide. The pumping energy is fed into a normal mode propagating in the waveguide and it is continuously absorbed by the maser material while saturating its pumping transition. The signal propagates in another normal mode of the waveguide, and, while continuously interacting with the maser material, it gains exponentially in amplitude by amplification through stimulated emission.

For high gain per unit length of the waveguide the signal wave should travel slowly so that it will have more time to interact with the maser material. The group velocity of the signal wave should therefore be low. This requirement corresponds to the requirement for high Q and resonance enhancement of fields in cavity masers.

There are a number of waveguide structures for microwaves which have slow modes of propagation. They are called microwave delay lines. Many such lines with particularly large delays have a periodic structure. We will therefore develop a general theory of the travelling wave maser for periodic delay lines. The more elementary theory for a uniform waveguide may be obtained from the general case by a limiting process.

We assume the periodic delay line to have fundamental sections of periodic length ΔL as shown in Fig. 49. As in the case for a uniform waveguide there are also normal modes of propagation in a periodn waveguide. Any one of these modes will differ in its field distributio

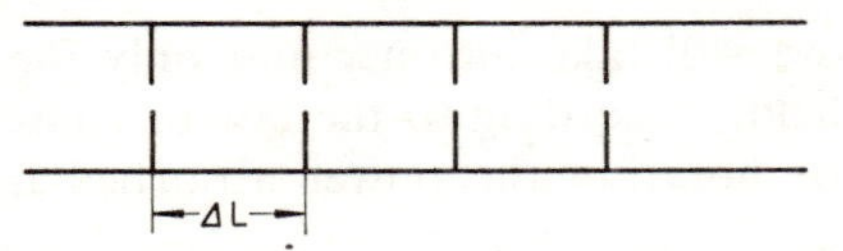

FIG. 49. Periodic delay line of a travelling wave maser

from period to period only by a constant factor e^{-g}. The exponent $g = a+jb$ is called the propagation factor. The gain constant v in an amplifying periodic structure follows from the real part of g:

$$v\Delta L = -a$$

To calculate the gain constant we define a quality factor Q for the periodic structure. As in case of the quality factor of resonator modes a quality factor may also be defined for the mode of propagation in a fundamental section of the periodic structure according to

$$Q = \omega \frac{\Delta W}{\Delta P} \tag{229}$$

Here ΔW represents the time average of the energy which the propagating mode stores within the fundamental section ΔL. ΔP is the time average of the power which the mode loses through absorption in the elementary section. In case of induced emission energy will be added to the signal mode ΔP and Q may then be negative.

The quantities ΔW and ΔP are determined from the field distribution of the respective propagating modes. The energy density of an a.c. magnetic field is $\frac{1}{2}\,\mathrm{Re}\,\mathbf{H}^*\cdot\mathbf{B}$. The magnetic field within the section will therefore store magnetic energy according to:

$$\Delta W_m = \tfrac{1}{2}\,\mathrm{Re}\iiint_{\Delta V} \mathbf{H}^*\cdot\mathbf{B}\,dV$$

This magnetic field energy ΔW_m of a mode of propagation equals the corresponding electric field energy since otherwise the energy of a standing wave in a waveguide resonator could not be completely exchanged between the electric and magnetic field during each half cycle. The time average of total energy within a section therefore is obtained as

$$\Delta W = 2\Delta W_m = \mathrm{Re} \iiint_{\Delta V} \mathbf{H}^* \cdot \mathbf{B}\, dV$$

For ΔP in (229) we will take into account only the power converted by the magnetic field. According to the law of conservation of energy in complex phasor notation the power absorbed from the magnetic field is

$$\Delta P = -\omega\, \mathrm{Im} \iiint_{\Delta V} \mathbf{H}^* \cdot \mathbf{B}\, dV$$

If the section were uniformly filled with an isotropic material of magnetic susceptibility $\chi = \chi' - j\chi''$ we would simply have

$$Q = \frac{1}{\chi''} \tag{230}$$

for the quality factor defined by (229). This relation between Q and χ' may now also be used to define a scalar magnetic susceptibility loss component χ''_{eff} for the general case of a section which is non-uniformly or only partially filled with an anisotropic material having a magnetic susceptibility tensor $\boldsymbol{\chi}$.

Following (230) and with the integral expressions for ΔW and ΔP substituted into (229) we obtain

$$\chi''_{\text{eff}} = \frac{1}{Q} = \frac{\iiint \mathbf{H}^+ \boldsymbol{\chi}'' \mathbf{H}\, dV}{\iiint |H|^2\, dV} \tag{231}$$

Note that the loss component $\boldsymbol{\chi}''$ of the susceptibility tensor obtains from

$$\boldsymbol{\chi}'' = \frac{1}{2j}(\boldsymbol{\chi} - \boldsymbol{\chi}^+)$$

where $\boldsymbol{\chi}^+$ is the Hermitian conjugate of $\boldsymbol{\chi}$ rather than the ordinary

$g(f)$ in the effective susceptibility χ''_{eff}, the gain depends on frequency. Delay lines can be designed to have a dispersion characteristic much less pronounced than the sharp frequency characteristic of the emission line. Hence for all practical purposes $g(f)$ determines the gain frequency characteristic. In turn the solid-state laser emission line is broadened by thermal relaxation times and hence forms a Gaussian curve.

Power gain of the travelling wave maser with matched impedances at input and output is given by the same expression (96) as for the laser amplifier power gain. The relation for maximum power gain at the centre frequency of the emission band is also the same as for laser amplifiers. The power gain bandwidth in the travelling wave maser narrows as the maximum gain at band centre increases. This effect was also noted for the laser amplifier and is true for any travelling wave amplifier. The 3-db bandwidth of power gain with a Lorentz curve for the emission line again follows from the same relation (97) as before for lasers. This bandwidth may also be read from the corresponding curve in Fig. 8. The gain bandwidth product of the travelling wave maser instead of remaining constant, as is the case for the cavity maser, grows with increasing gain. This is an important factor in making the travelling wave maser superior to the cavity maser.

As a further advantage the travelling wave maser is much more stable and less sensitive to fluctuation in the emission characteristics. Stable operation of a travelling wave maser is possible under normal conditions because the interaction between waves and spins, being non-reciprocal, can have high gain in the forward direction while, in the reverse direction, there is isolation or attenuation. This non-reciprocal behaviour shows up in the imaginary component of the susceptibility tensor. In general this tensor is not symmetric about the diagonal and, for isolated electron spins as well as paramagnetic crystals with axial symmetry, the imaginary component of χ may even be antisymmetric.

With such antisymmetry waves of the proper circular polarization for the magnetic field will have maximum gain due to stimulated emission while waves with the opposite sense of circular polarization will have no gain and instead be attenuated through absorption losses. The delay line may be designed for the forward signal wave to have the proper polarization for maximum gain and the backward signal wave with the opposite polarization to be attenuated. In the last section of this chapter we will discuss an actual delay line which displays these characteristics. Some practical travelling wave maser arrangements also employ ferrite loading of the delay line to obtain non-reciprocal loss through ferromagnetic resonance. Travelling wave masers are normally designed to be unconditionally stable. They may be terminated by arbitrary load impedances at the input and output without the self-excitation of oscillations.

Travelling wave masers in general also have a larger dynamic range for gain than cavity masers. They will amplify linearly up to higher output powers before they go into saturation. The cavity maser has a saturated output power which is proportional to the volume of maser material in the cavity. From a certain power level the maser material will go into saturation uniformly throughout the cavity. This uniform saturation begins when, with increasing stimulated emission, the transition rate w in eqn. (227) becomes as high as the overall rate $s+s'$ of thermal transitions due to relaxation phenomena. With the travelling wave maser the active material will likewise go into saturation when $w = s+s'$. But it will not begin to saturate uniformly. The end of the delay line near the amplifier output will be saturated first. With the input power growing still further this range of saturation will then extend further and further towards the input end of the delay line. The output power still grows during this phase of spreading saturation, but not linearly with input power. Maximum output power will only then be obtained when the maser material is completely saturated all along the delay line. This maximum output power is, therefore, also proportional to the volume of the maser material as is the case for the cavity maser. The travelling wave maser may, however, accommodate much more maser material over all of its delay line length than the resonator cavity of a cavity maser. The saturated output power will therefore always be larger in the travelling wave maser.

3.6. Noise in the travelling wave maser

Travelling wave masers as well as most other types of maser have their most important application as low noise amplifiers. Even without any special measures maser noise is quite low. In addition for all applications where extremely low noise is required masers must be designed to utilize the intrinsically low noise and actually yield high sensitivity in amplification. For masers, as well as for lasers earlier in our discussion, not only must the usual thermal noise be taken into account but, in addition, noise due to spontaneous emission is of significance. For the laser and the very high frequencies which it generates or amplifies, individual photons have so much energy that the noise of spontaneous emission is much higher than thermal noise. In masers these two noise contributions are in a quite different ratio. Here the thermal noise is normally much larger. But, in the ultimate limit, we must also take into account the noise due to spontaneous emission.

First we will calculate the output power due to thermal noise in the travelling wave maser. To obtain the thermal noise we have to consider the attenuation α which the signal wave will experience along the delay line in the absence of induced emission. Thermal noise will only be generated when the delay line has attenuation. Lossless or absorptionless structures do not absorb thermal radiation and hence will also not radiate any thermal noise. The rate of change of thermal noise power P along a transmission line follows from

$$\frac{dP}{dz} = 2vP - 2\alpha P + R \tag{234}$$

Here $2vP$ represents the increase in thermal noise power per unit axial length due to induced emission and $2\alpha P$ is the power lost per unit length due to transmission line attenuation. R designates the power increase per unit length due to thermal noise. We will determine R by suppressing induced emission. For a passive delay line we have $2v = 0$ and through any cross-section the same amount of thermal noise power

$$P_0 = \frac{hf}{e^{\frac{hf}{kT_M}} - 1}\,\delta f \tag{235}$$

is transmitted. According to eqn. (109) P_0 is the radiation generated in thermal equilibrium by any thermal noise source at absolute temperature T_M within the frequency range δf. T_M designates the temperature of the maser which, for a cooled system, differs from the room temperature T of the input source. Because the power P_0 is the same for any cross-section we have $dP/dz = 0$ in (234) and hence

$$-2\alpha P_0 + R = 0$$

or

$$R = 2\alpha P_0$$

Returning to the normal operating conditions with induced emission eqn. (239) then reads

$$\frac{dP}{dz} = 2(v-\alpha)P + 2\alpha P_0 \tag{236}$$

This is the same form of the differential equation as (105) for the noise due to spontaneous emission in lasers. Only v must be replaced by $v-\alpha$ and R_0 by $2\alpha P_0$ to obtain (236) from (105). The solution of this differential equation will, therefore, also have the same form as the solution (106) of (105). Hence the thermal noise power at the output of the maser is given by

$$P = \frac{\alpha P_0}{v-\alpha}(e^{2(v-\alpha)L}-1) \tag{237}$$

We will now generalize this expression by accounting also for the noise contributed by spontaneous emission. The same considerations which resulted in (107) for the spontaneous noise of the laser will also lead to a corresponding expression here. Except that under the present circumstances of delay line attenuation the spontaneous emission will not be amplified according to the gain constant v, but only according to $v-\alpha$. Instead of (107) we have therefore the following expression for the noise power of spontaneous emission in travelling wave masers:

$$P_R = \frac{v}{v-\alpha}\,\frac{n_2}{n_2-n_1}\,hf_{12}\delta f(e^{2(v-\alpha)L}-1) \tag{238}$$

n_2 and n_1 still designate the density of states of the upper and lower levels of induced emission.

As a characteristic quantity to describe the noise of a maser we use the noise figure according to its normal definition:

$$F = \frac{P_T+P+P_R}{P_T}$$

If we now substitute from (235), (237) and (238) and, as in (96), designate the power gain by V, the following expression results for the noise figure

$$F = 1+\frac{v}{v-\alpha}\left(1-\frac{1}{V}\right)\left[\frac{\alpha}{v}\,\frac{1}{e^{\frac{hf}{kT_M}}-1}+\frac{n_2}{n_2-n_1}\right]\left(e^{\frac{hf}{kT}}-1\right) \tag{239}$$

When the noise due to spontaneous emission in solid state masers is considered it is of some advantage to introduce the concept of negative spin temperature. In thermal equilibrium at absolute temperature T without any pump radiation applied the various states are populated according to Boltzmann statistics as follows:

$$\frac{n_1}{n_2} = e^{\frac{hf}{kT}}$$

This relation is retained even now when the population of states is inverted by pumping radiation and the temperature T becomes negative. A temperature T_s is defined by

$$\frac{n_1}{n_2} = e^{\frac{hf}{kT_S}}$$

for any population distribution including also non-equilibrium distributions which do not then correspond to Boltzmann statistics. For $n_1 = n_2$ this temperature grows without limit ($T_S \to \infty$) while for $n_1 < n_2$ it becomes negative ($T_S < 0$).

The closely spaced energy levels of maser transitions normally have only a weak inversion density. In this case

$$\frac{n_1}{n_2} \simeq 1 + \frac{hf}{kT_S} \tag{240}$$

or

$$T_S \simeq \frac{hf}{k} \frac{n_2}{n_1 - n_2} \tag{241}$$

For such a low inversion density its ratio with respect to n_2 follows from (226) as

$$\frac{n_1 - n_2}{n_2} = -\frac{3\Delta N}{N} = -\frac{hf}{kT_M}\left(\frac{s'}{s'+s}\frac{f_p}{f} - 1\right) \tag{242}$$

It has been assumed here that $w = 0$. When we need to consider the sensitivity of an amplifier its input power will always be so low that the rate of transitions induced by this input power is very low also.

Substituting from (242) into (241) the spin temperature under these low input conditions results in

$$T_S \simeq -\frac{T_M}{\frac{s'}{s'+s}\frac{f_p}{f} - 1} \tag{243}$$

The expression (239) for the noise figure may also be simplified when we have, for frequencies in the microwave range, $hf \ll kT$ and, as in most cases, $hf \ll kT_M$ also.

If in addition we express the relative inversion density by the corresponding spin temperature according to (241) we obtain

$$F = 1+\frac{v}{v-\alpha}\left(1-\frac{1}{V}\right)\left(\frac{\alpha}{v}\frac{T_M}{T}-\frac{T_S}{T}\right) \tag{244}$$

In the limiting case of high gain ($v \gg \alpha$ as well as $V \gg 1$) and equal relaxation times for both thermal transitions ($s' = s$) the noise figure reduces to

$$F = 1+\left(\frac{\alpha}{v}+\frac{1}{\frac{f_p}{2f}-1}\right)\frac{T_M}{T} \tag{245}$$

To obtain high sensitivity from this relation the attenuation of the passive maser structure should be as low as possible while the frequency f_p of the pumping transition should be large compared with the frequency f of the signal transition. Under present conditions the equivalent noise temperature of the maser follows from (245) as

$$T_R = \left(\frac{\alpha}{v}+\frac{1}{\frac{f_p}{2f}-1}\right)T_M \tag{246}$$

Only if $\alpha \ll v$ and $fp \gg 2f$ does this noise temperature drop to values much lower than the maser temperature T_M. But T_M by itself is normally quite low because of cryogenic cooling.

3.7. Travelling wave maser with comb line

Of all delay line structures the comb line has proven most suitable for maser amplifiers. Nearly all travelling wave masers are therefore built using comb structures for the delay line. As shown in Fig. 50, a comb line consists of a metallic tube with rectangular cross-section having a comb-like arrangement of metallic fingers. These comb fingers extend in a transverse direction and are attached to one of the side walls.

In the direction of the fingers the comb forms a multi-conductor transmission line with as many conductors as the comb has fingers plus one for the top and bottom walls of the tube. One end of the multi-conductor transmission line is short-circuited by the side wall, the other end is terminated by the stray capacitances C_S between the tip of each finger and the other side wall of the rectangular tube.

To find the normal modes of propagation on this comb structure we assume it to extend to infinity in the longitudinal direction. In the transverse direction we then have a multi-conductor transmission line consisting of an infinite number of conductors.

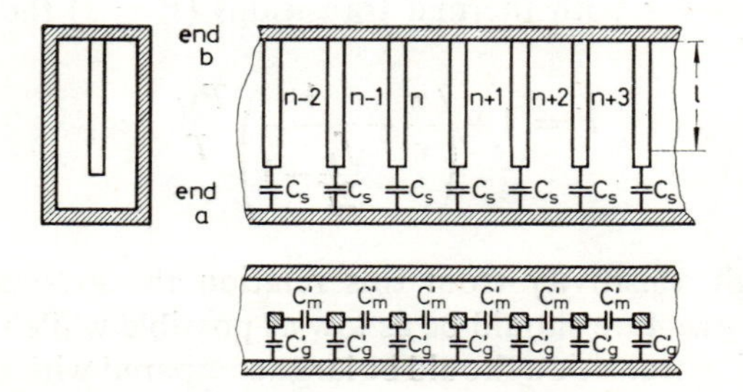

FIG. 50. Comb line with end capacitances C_s and distributed capacitances C'_m and C'_g per unit length of the comb fingers

The transmission line equations for TEM waves in the transverse direction may be obtained from the general theory of multi-conductor lines [34]. Solved for the voltages at each end of the multiconductor line these equations are, in matrix notation,

$$c\mathbf{K}\mathbf{V}_a = -\frac{j}{\tan kl}\,\mathbf{I}_a + \frac{j}{\sin kl}\,\mathbf{I}_b \tag{247}$$

$$c\mathbf{K}\mathbf{V}_b = \frac{j}{\tan kl}\,\mathbf{I}_b - \frac{j}{\sin kl}\,\mathbf{I}_a \tag{248}$$

Here $\mathbf{I}_a$ and $\mathbf{V}_a$ and $\mathbf{I}_b$ and $\mathbf{V}_b$ are column vectors of currents and voltages with respect to the top and bottom walls at both ends of the fingers, where the top and bottom walls are at ground potential. The elements of these column vectors are the phasors of the corresponding end currents and end voltages. $k = \omega\sqrt{\mu\varepsilon}$ is the wave number and $c = 1/\sqrt{\mu\varepsilon}$ the velocity of propagation for uniform plane waves in the medium with permeability μ and permittivity ε between the conductors. $\mathbf{K}$ is the matrix of distributed capacitances for the multi-conductor line. Accordingly

$$\mathbf{K} = \frac{1}{j\omega}\,\mathbf{Y}$$

gives the matrix **Y** of shunt admittances which are distributed along the line per unit length. In Fig. 50 each finger of the comb has been assumed to have distributed capacitances to the top and bottom walls, denoted by C'_g, and in addition to both neighbouring conductors, denoted by C'_m. The capacitance between a particular conductor and all other conductors except the immediate neighbours may be neglected. Such remote conductors are fairly well shielded from each other. With this assumption the matrix of distributed capacitances reduces to:

$$\mathbf{K} = \begin{bmatrix} \cdots & \cdot & \cdot & \cdot & \cdots \\ \cdots & -C'_m & 0 & 0 & \cdots \\ \cdots & C'_g+2C'_m & -C'_m & 0 & \cdots \\ \cdots & -C'_m & C'_g+2C'_m & -C'_m & \cdots \\ \cdots & 0 & -C'_m & C'_g+2C'_m & \cdots \\ \cdots & 0 & 0 & -C'_m & \cdots \\ \cdots & \cdot & \cdot & \cdot & \cdots \end{bmatrix} \tag{249}$$

Only the elements on the main diagonal and their immediate neighbours are different from zero.

At the end b in Fig. 50 all fingers are shorted by the side wall, hence

$$\mathbf{V}_b = 0$$

At the other end a they are terminated by capacitances C_S, therefore

$$\mathbf{I}_a = -j\omega C_S \mathbf{V}_a \tag{250}$$

With these terminal conditions we obtain from (248)

$$\mathbf{I}_b = \frac{1}{\cos kl}\mathbf{I}_a \tag{251}$$

while from (247) we obtain

$$c\mathbf{K}\mathbf{V}_a = \omega C_S \tan kl\, \mathbf{V}_a$$

The latter expression represents a system of equations for the voltages at the tip of each finger. For the nth finger and its neighbours it follows from this system that

$$-C'_m V_{a(n-1)} + (C'_g + 2C'_m)V_{a(n)} - C'_m V_{a(n+1)} = kC_S \tan kl V_{a(n)} \tag{252}$$

This is a linear and homogeneous difference equation of second order.

For its solution we let

$$V_{a(n)} = V_{a(n-1)}e^{-j\varphi} \tag{253}$$

and, by substituting into (252), obtain the following characteristic equation for the phase shift φ between voltages at neighbouring finger tips

$$\cos\varphi = 1+\frac{C_g'}{2C_m'}-\frac{C_S}{2C_m'l}kl\tan kl \tag{254}$$

Values of φ which solve this equation represent the phase factor of normal modes on the comb line. Figure 51 shows the graphical solution of (254). In Fig. 51a the right-hand side of (254) has been plotted versus

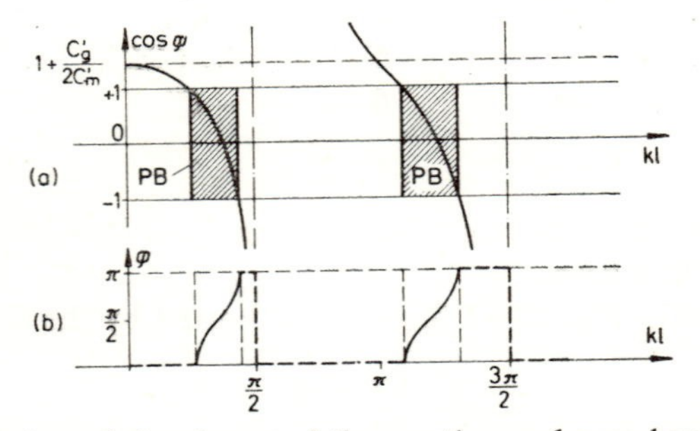

FIG. 51. Solution of the characteristic equation and pass band PB of the comb line

kl for a fixed value of l thus giving a represention as a function of frequency. Solutions with real φ require $\cos\varphi \ll 1$. These are pass bands (PB) of the comb line. Figure 51b shows φ as a function of kl within these pass bands. Travelling wave masers normally make use of the first pass band at $kl \ll \pi/2$. To obtain a large group delay $t_g = d\varphi/d\omega$ the phase characteristic should be very steep, thus requiring a very narrow pass band. The fingers of a comb line are designed and arranged to meet this requirement.

Once the phase factor has been determined the distribution of voltages and currents along the comb line may be found from (250), (251), and (253). These voltages and currents will also give an idea of the field distribution.

At the lower limit of the first pass band we have $\varphi = 0$. The voltages at all finger tips are in phase. The electric field has, therefore, the same phase everywhere along the line and is distributed as shown in Fig. 52a in a longitudinal section. The wavelength of propagation under these conditions is infinitely large.

At the upper limit of the first pass band we have $\varphi = \pi$. Now the voltages of neighbouring fingers are in phase opposition. The electric field is distributed as shown in Fig. 52b. Except for the oppositely directed field lines from one finger to the next, the field is of equal phase all along the comb line in this case also. The wavelength of propagation corresponds to two periods of the comb line. Within the pass band the

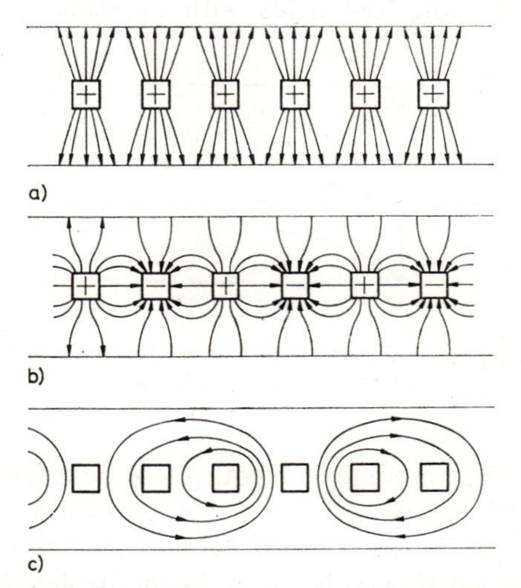

FIG. 52. Typical field distribution in the first pass band of the comb line: (a) Electric field for the lower cut-off frequency. (b) Electric field for the upper cut-off frequency. (c) Magnetic field for a medium pass-band frequency

phase factor lies between 0 and π and the wavelength extends over more then two sections of the comb structure. Except for the stray fields between the finger tips of the comb and the adjacent side wall all electric field lines as well as the magnetic field lines lie in longitudinal sections parallel to the side walls. This purely transverse character of both fields with respect to the comb fingers justifies the analysis of the TEM waves on a multiconductor line.

For a frequency somewhere in the middle of the first pass band the lines of the instantaneous magnetic field run as shown in Fig. 52c. As time progresses this instantaneous field moves along the comb line with the phase velocity $v_p = \omega \Delta L/\varphi$. The field pattern remains unchanged except for the immediate vicinity of the comb fingers where the boundary conditions at the metallic surfaces must be continuously satisfied. For an observer in a fixed position somewhat below and somewhat above the comb in Fig. 52c the magnetic field vector rotates clockwise or counter-clockwise, respectively, as the field pattern moves along. At these locations the magnetic field is therefore nearly circularly polarized.

In general the magnetic field has an elliptic polarization in the plane of a longitudinal section. This elliptical polarization may be separated into two circularly polarized fields with opposite sense of rotation.

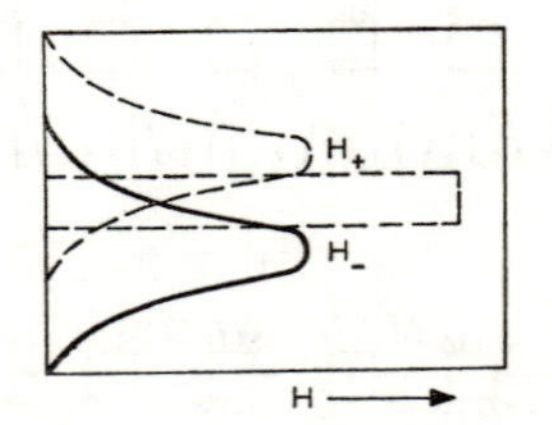

Fig. 53. Circularly polarized components of the magnetic field in longitudinal sections of the comb line

Figure 53 shows the amplitude of both these polarizations versus the vertical cross-sectional coordinate for a cross-section between two comb fingers. Right in the centre, between the comb fingers, both polarizations have equal magnitude, and here the field is linearly polarized. Upwards or downwards from this centre one of the polarizations decreases, reaches a maximum and then decreases further until it vanishes at the top or bottom wall.

For the largest interaction of this magnetic field with a material which has isolated spin states the material should be placed in the form of a strip either above or below the comb where one of the circular polarizations has a maximum amplitude. This strip must then be magnetized transverse to the longitudinal section of Fig. 52c. The comb wave will be amplified by this arrangement only when it travels in one direction. When travelling in the opposite direction the polarization also has the opposite rotation and will not interact significantly with the spin states.

For the spin states of paramagnetic crystals the susceptibility tensor does not have the simple form (207) of isolated spin states. Here the more general σ-vector must be considered. For ruby expression (211) must be applied, or, when we rotate the coordinate system around the y-axis by the angle (212), the more convenient form (213) holds. Efficient interaction between spin states and the r.f. field is obtained when the c-axis of the crystal is oriented at $\theta = 90°$ with respect to the direction of the magnetization. Under these circumstances we have $\alpha \ll \gamma$ in (210) for all transitions between neighbouring spin states. The orientation of the z'-axis will then also be at

$$\phi = 90°$$

with respect to the c-axis of the crystal. The crystal must then be magnetized parallel to the z'-axis.

From (213) we obtain

$$\sigma\sigma^{+} = \frac{1}{4}\begin{bmatrix} \gamma^2 & -j\beta\gamma & 0 \\ j\beta\gamma & \beta^2 & 0 \\ 0 & 0 & 0 \end{bmatrix}$$

as a good approximation for the susceptibility tensor.

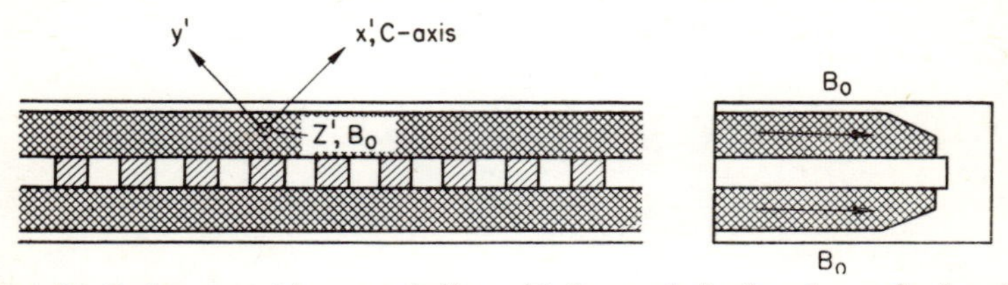

FIG. 54. Ruby crystal in a comb line with its c-axis in the plane of a longitudinal section

Under these circumstances transitions are induced only by those a.c. magnetic fields which have a circular polarization in the $x'-y'$ plane. The crystal is therefore placed in the comb structure so that its c-axis lies in the plane of a longitudinal section as shown in Fig. 54 [35]. The d.c. magnetization is applied parallel to the comb fingers. The situation is then nearly identical to that for isolated spin states and no crystalline anisotropy.

The dielectric properties of the ruby insert will modify the field distribution of the comb wave. However the basic polarization characteristics still prevail. Only when the group delay is to be calculated from φ and the effective susceptibility from (231) must we consider the dielectric characteristics of the ruby insert.

In the microwave range ruby has a relative permittivity of $\varepsilon_r = 9$. To obtain a high negative χ''_{eff} and, with it, large gain, almost the whole cross-section between the comb and top and bottom walls is filled with ruby (Fig. 54). The high permittivity of the insert will raise the group delay and thus also enhance the gain. The advantage of higher gain through more group delay is sufficient to forego a non-reciprocal gain obtained with a ruby insert on only one side. Instead the backward gain for stable operation is obtained by additional inserts of ferromagnetic material. With the same magnetization as for the ruby these inserts are magnetized to absorb the backward comb wave through ferromagnetic resonance absorption. To apply pump power to the maser crystal at the much higher frequency f_p the power is launched into a higher order mode of the rectangular tube and comb structure. This mode corresponds to the dominant TE_{10} wave of the empty rectangular tube as modified by the comb and maser material.

Bibliography

Monographs and review papers

1. A. A. VUYLSTEKE, *Elements of Maser Theory*. D. van Nostrand, New York (1960).
2. G. TROUP, *Masers and Lasers*. Methuen, London (1963).
3. M. BROTHERTON, *Masers and Lasers*. McGraw-Hill, New York (1964).
4. W. H. LOUISELL, *Radiation and Noise in Quantum Electronics*. McGraw-Hill, New York (1964).
5. D. RÖSS, *Laser Lichtverstärker und -oszillatoren*. Akadem. Verlagsgesellschaft, Frankfurt/M. (1966).
6. G. BIRNBAUM, *Optical Masers*. Academic Press, New York (1964).
7. B. A. LENGYEL, *Lasers*. John Wiley, New York (1963).
8. B. A. LENGYEL, *Introduction to Laser Physics*. John Wiley, New York (1966).
9. O. S. HEAVENS, *Optical Masers*. Methuen, London (1964).
10. A. YARIV and J. P. GORDON, The laser. *Proc. IEEE* **51,** 4–29 (1963).
11. R. SALTONSTALL, *Laser Technology*. Hobbs, Dorman & Co., New York (1965).
12. M. ROSS, *Laser Receivers*. John Wiley, New York (1966).
13. J. R. SINGER, *Masers*. John Wiley, New York (1959).
14. A. E. SIEGMAN, *Microwave Solid-state Masers*. McGraw-Hill, New York (1964).
15. *Applied Optics*, Vol. 5, No. 10, October 1966.
    Proc. IEEE, Vol. 54, No. 10, October 1966.

15a. V. M. FAIN and YA. I. KHANIN, *Quantum Electronics*, Vols. 1 and 2. Pergamon Press (1969).

References

The following list of references is by no means complete. Compiled here are only those articles containing particular results which are quoted in the preceding text without further explanation or derivation. A fairly complete list of references on lasers may be found in ref. 5 and on solid state masers in ref. 14. In addition the following list of references quotes the sources for analytical methods which are used in the preceding text.

16. R. F. HARRINGSTON, *Time-Harmonic Electromagnetic Fields.* McGraw-Hill, New York (1961).
17. G. D. BOYD and H. KOGELNIK, Generalized Confocal Resonator Theory. *Bell Syst. Techn. J.* **41,** 1347–1369 (1962).
18. M. ABRAMOWITZ and J. A. STEGUM, *Handbook of Mathematical Functions.* N.B.S. Washington (1964).
19. G. GOUBAU and F. SCHWERING, On the propagation of electromagnetic wave beams. *Trans IRE* **AP–9,** 248–256 (1961).
20. G. D. BOYD and J. P. GORDON, Confocal multimode resonators for millimeter through optical wavelength masers. *Bell Syst. Techn. J.* **40,** 489–508 (1961).
21. C. FLAMMER, *Spheroidal Wave Functions.* Stanford Univ. Press, Stanford (1957).
22. H. KOGELNIK, *Modes in Optical Resonators in Lasers,* A. K. LEVINE. Dekker, New York (1966).
23. H. KOGELNIK and T. LI, Laser beams and resonators. *Proc. IEEE* **54,** 1312–1329 (1966).
24. D. GLOGE, Ein allgemeines Verfahren zur Berechnung optischer Resonatoren und periodischer Linsensysteme. *A.E.Ü.* **19,** 13–26 (1965).
25. A. G. FOX and T. LI, Resonant modes in a maser interferometer. *Bell Syst. Techn. J.* **40,** 453–488 (1961).
26. W. HEITMANN, Zinkselenid als hochbrechende Substanz in dielektrischen Spiegelschichten für Laser-Resonatoren und Interferenzfilter. *Z. Angew. Phys.* **19,** 392–395 (1965).
27. D. L. PERRY, Broadband dielectric mirrors for multiple wavelength laser operation in the visible. *Proc. IEEE* **53,** 76–82 (1965).
28. D. TER HAAR, *Elements of Statistical Mechanics.* Holt, Rinehart & Winston, New York (1961).
29. H. KOGELNIK and C. K. N. PATEL, Mode suppressing and single frequency operation in gaseous optical masers. *Proc. IRE* **50,** 2365–2366 (1962).
30. E. I. GORDON and A. D. WHITE, Single frequency gas lasers at 6328 Å. *Proc. IEEE* **52,** 206–207 (1964).

31. D. Röss, Room-temperature cw ruby laser. *Microwaves* **3,** 29–33 (1965).
32. A. E. Siegman, Gain bandwidth and noise in maser amplifiers. *Proc. IEEE* **45,** 1737–1738 (1957).
33. M. W. P. Strandberg, Unidirectional paramagnetic amplifier design. *Proc. IRE* **48,** 1307–1320 (1960).
34. H. J. von Baeyer and R. Knechtli, Über die Behandlung von Mehrleitersystemen mit TEM-Wellen bei hohen Frequenzen. *Zeitschr. f. angew. Mathematik und Physik,* **3,** 271–286 (1952).
35. W. J. Tabor and J. T. Sibilia, Masers for the Telstar satellite communications experiment. *Bell Syst. Techn. J.* **42,** 1863–1886 (1963).

Problems

1. Derive eqn. (3) by assuming that the forces which tie the electron to the potential well are much larger than the external forces.
2. An electron is moving on a circular orbit of radius r around the nucleus of a hydrogen atom. Let the mass of the nucleus $m_n \to \infty$. Find the Lagrange and Hamilton functions L and H for this system. Derive the general coordinates of momentum and the canonical equations of motion. Show that L and H describe the electron motion correctly.
3. A point of mass m moves on a circular orbit of radius r in the plane $z = \text{const.}$ For this case derive eqns. (6), (7) and (10) in cylindrical coordinates.
4. Find the vector potential $\mathbf{A}$ with the scalar potential $\varphi = 0$ for a plane uniform wave which is linearly polarized in x-direction and which has a real Poynting vector $\mathbf{S} = \mathbf{u}_z S_z$.
5. Show that the momentum operator $\bar{\mathbf{p}} = -j\hbar\nabla$ in spherical coordinates satisfies the commutator relation $[q_i, p_k] = j\hbar\delta_{ik}$.
6. Derive the time-independent Schrödinger equation for the hydrogen atom with a nucleus of mass $m_n \to \infty$.
   (a) Which is the form of this equation when only eigenfunctions of spherical symmetry are considered?
   (b) The asymptotic form of these eigenfunctions ψ_{as} for $r \to \infty$ is to be calculated. Assume here that $\dfrac{2mW}{\hbar^2} = -\dfrac{1}{4r_0^2}$ so that we have $W < 0$. Which of the two possible solutions is physically meaningful?
   (c) In order to calculate the eigenfunctions $\psi(\varrho)$ with $\varrho = \dfrac{r}{r_0}$ one lets $\psi(\varrho) = \psi_{as}v(\varrho)$. For $\psi(\varrho)$ to behave properly at $\varrho \to \infty$ and its square to be integrable, $v(\varrho)$ must be a polynomial of finite order. What is the differential equation for $v(\varrho)$?
   (d) The Laguerre differential equation $\varrho y'' + (1-\varrho)y' + ky = 0$ is solved for an integral $k \geqslant 0$ by the Laguerre polynominal $L_k(\varrho)$. If k is not an integer the solutions are infinite series. The Laguerre polynomials are defined by

$$L_k(\varrho) = \frac{1}{k!}\, e^{\varrho}\, \frac{d^k}{d\varrho^k}\,(\varrho^k e^{-\varrho})$$

   If the Laguerre differential equation is differentiated with respect to ϱ an equation is obtained for $L_k^{(1)}(\varrho) = \dfrac{dL_k(\varrho)}{d\varrho}$ which is identical with the differential equation for $v(\varrho)$. By comparing coefficients find the eigenfunctions of spherical symmetry and their eigenvalues.

7. Show that the eigenfunctions of the time-independent Schrödinger equation are orthogonal. To this end multiply the equation $\bar{H}\psi_n = W_n\psi_n$ by ψ_m^* and the complex conjugate of $\bar{H}\psi_m = W_m\psi_m$ by ψ_n. Subtract one of the resulting equations from the other and integrate over the full range of all position coordinates. If the boundary conditions for ψ at $r \to \infty$ are taken into account, the resulting integral may be shown to vanish.
8. Prove that the Hermitian operators always have real eigenvalues by showing that the eigenvalue λ of the corresponding eigenvalue equation is equal to the eigenvalue λ^* of the complex conjugate eigenvalue equation.
9. Let φ_1 be an eigenfunction of the operator $\bar{G}_1$ and let $\bar{G}_1$ commute with $\bar{G}_2$. By evaluating $\bar{G}_1\bar{G}_2\varphi_1$ show that φ_1 is also an eigenfunction of $\bar{G}_2$, meaning that operators which commute have identical sets of eigenfunctions.
10. A micro-system is in state 2 for $t \leqslant 0$. At $t = 0$ let a perturbation with $H_{12}(t) = c\delta(t)$ act upon the system where $\delta(t)$ is Dirac's δ-function. What is the probability $w_{12}(t)$ for a transition to state 1 at $t > 0$?
11. For a more precise determination of $a_m(t)$ in eqn. (30) calculate the second iteration.
12. Apply the perturbation operator $\bar{H}_t$ in (36) to the ground state of the hydrogen atom with state function $\psi_1(r) = C_1 e^{-\frac{r}{2r_0}}$. Compare the term which is linear in **A** to the quadratic **A**-term. For a plane wave with wavelength $\lambda = 1\ \mu$m specify the maximum electric field for which the quadratic term may still be neglected.
13. Use the eigenfunctions of spherical symmetry and low order for the hydrogen atom to show that for optical frequencies the electromagnetic field $\mathbf{A}_i = \mathbf{A}(\mathbf{r}_i, t)$ may be represented by its value $\mathbf{A}_0 = \mathbf{A}(0, t)$ at the centre of the atom over the extension of the micro-system.
14. Use the hydrogen atom to show that the matrix element of the electric dipole moment is zero for all states of spherical symmetry and that hence there are no electric dipole transitions between such states.
15. Equation (48) is to be derived.
16. Discuss the transition probability (51) as a function of time and frequency difference $\Delta\omega = \omega - \omega_{nm}$. Specify all frequencies at which $w_{nm}(\Delta\omega)$ assumes maximum values.
17. Show from the classical model that

$$P = \frac{e^2 r_0^2 \omega^4}{12\pi\varepsilon c^3}$$

is the power which an electron radiates when oscillating according to $r = r_0 \sin \omega t$.

18. Derive the separation equation (64) from the wave equation.
19. Let an emission line be broadened by finite lifetime only. What is the difference δf between the transition frequency f_{12} and the frequency of an external oscillation for the excitation factor Λ to drop to 10 per cent of its maximum value?
20. Evaluate $\int_0^\infty g(f)df$ for a Lorentzian shape of the emission line.

21. A three level laser medium with n active micro-systems per cm^3 at therma equilibrium is irradiated by linearly-polarized light of spectral intensity $I(f) = I_0 g(f)$. The medium extends over the length L in the direction of the propagation for the incident light. After passing through the medium the light intensity is measured at I_L. The intensity has decreased due to induced absorption between states 1 and 2. The transition has a normalized line shape $g_2(f)$. The transition frequency is f_{12}. Find B_{12}, $|P_{12}|_E$ and A_{12} by assuming that always $\alpha L \ll 1$, and by neglecting reflections at the end faces of the medium.

22. For a laser oscillator with an inhomogeneously broadened emission-line for the gain constant

$$v = v_0 e^{-\ln 2 \left(2 \frac{f-f_0}{\Delta f}\right)^2}$$

and with resonator losses α, what should the spacing L between mirrors be so that only one of the lowest transverse order modes is excited?

23. Find the equivalent input noise P_{RE} and the noise temperature of spontaneous emission for arbitrary gain and inversion of states.

24. Formulate the rate equations of a four-level laser with a lower state 1 of induced transitions discharging to the ground state with the relaxation time τ_{01}. Calculate N_1/N_2 and N_2 and $\Delta N = N_1 - N_2$ as a function of the number p of photons for only one mode oscillating in the resonator from the stationary solution of these rate equations.

25. The integral equation (160)

$$u(x_2) = \frac{\varkappa_x}{\sqrt{\lambda L}} \int_{+\infty}^{-\infty} K(x_1, x_2) u(x_1)\, dx_1$$

with $K(x_1, x_2)$ according to (161) may be considered an operator equation $\bar{G}u(x_1) = Gu(x_2)$ with the operator

$$\bar{G} = \int_{-\infty}^{+\infty} dx_1 K(x_1, x_2)$$

and the eigenvalue

$$G = \frac{\sqrt{\lambda L}}{\varkappa_x}$$

Show that $\bar{G}$ commutes with the Hamilton operator of the linear harmonic oscillator $\bar{H} = \frac{d}{dx^2} - \alpha^4 x^2$ and therefore has the same set of eigenfunctions.

26. In order to derive eqn. (169) for the position of the phase front along the resonator axis. Let

$$u(x_1) = e^{-\left(\frac{x_1}{w_1}\right)^2} H_n\left(\frac{x_1}{w_1}\sqrt{2}\right)$$

be the amplitude distribution at $z = -L_1/2$ with R_1 the curvature radius of the phase front at $z = -L_1/2$ and evaluate the integral for the field amplitude at z

by using

$$\int_{-\infty}^{+\infty} e^{-(x-y)^2} H_n(\alpha x)\, dx = \sqrt{\pi}(1-\alpha^2)^{\frac{n}{2}} H_n\left[\frac{y}{(1-\alpha^2)^{\frac{1}{2}}}\right]$$

27. The general conditions (170) for stable modes in optical resonators may also be formulated as follows: For stable modes to exist either the centre of curvature of one mirror or the mirror itself but not both must be located between the other mirror and its centre of curvature. Show that these conditions are identical to (170).

28. For a confocal resonator with round mirrors $N = 1$ and $L/\lambda = 10^6$ find the mean lifetime of photons in a mode of lowest transverse order and the Q-factor and finesse of these modes. The finesse of an optical resonator is the ratio of the 3-db bandwidth of a mode to the frequency spacing between two adjacent modes of the same transverse order, but different longitudinal orders.

29. In a confocal optical resonator how many modes of one longitudinal order and higher transverse orders have resonant frequencies between two lowest transverse order modes of the same and the next higher longitudinal order?

30. Find the classical angular velocity ω_p of electron spin precession in a d.c. magnetic field $\mathbf{B}_0$.

31. Show that the matrices of the spin components anticommute according to

$$\mathbf{S}_i\mathbf{S}_j + \mathbf{S}_j\mathbf{S}_i = \frac{\hbar^2}{2}\,\delta_{ij}\mathbf{1}$$

32. Prove the identity $\quad \mathrm{Im}(\mathbf{H}^* B) = -\mu_0(\mathbf{H}^+ \boldsymbol{\chi}_m'' \mathbf{H})$

where

$$\boldsymbol{\chi}_m'' = \frac{1}{2j}[\boldsymbol{\chi}_m - \boldsymbol{\chi}_m^+]$$

Index

Index

Index

OTHER TITLES IN THE SERIES IN NATURAL PHILOSOPHY